T0093572

'Jake Robinson is the Gilbert White, the Henry David Thoreau, of the microbiome. His charmingly written book, a work of science leavened by literary allusion and engaging personal memoir, invites us to dive down through many powers of ten to the invisible level of microbes. Levels, rather, for "microbes" range in size from each other as much as we do from them. Microbes live in us, on us, through us, about us. Bacteria include the green photosynthesis specialists that inhabit the solar panels we call leaves, and are the ultimate source of all our food and oxygen. Each of our cells is an ecosystem of tiny biochemists, whose R and D we borrow to stay alive through every next second. Microbes invented antibiotics for their own protection, megayears before we hijacked them (and abused them) for ours. Micro-organisms are our foes, but our indispensable friends too. Do they even, as Robinson proposes, manipulate us and our behaviour to their, and our, benefit? Not just us but bumblebees and trees, and who knows what else? Do bacteria in clouds make rain, again to their advantage? Such suggestions, music to my ears, may prove controversial but nobody could fail to be intrigued.'

—Richard Dawkins

'I enjoyed this book very much indeed. It was a fascinating romp through the microbial world and is definitely the sort of book that would make me miss my station. If you are not a microbiologist, you will be astonished at how much microbes affect, regulate and sustain your life, and Robinson's passion and enthusiasm for the world of the very tiny positively glows on every page.'

—Dr George McGavin

'Robinson has crafted an immensely accessible and important book. It lifts the lid on the secrets of how our microbial world is so crucial to human and planetary health.'

—Prof. John F. Cryan, Principal Investigator in the APC Microbiome Institute, University College Cork

'*Invisible Friends* is beautifully written, and packed with information about the tiny organisms that form the basis of life on Earth. From the microbes in our bodies to those on the International Space Station, Jake Robinson reveals a hidden world that we would be unwise to ignore.'

—Rebecca Nesbit, ecologist and author of *Tickets for the Ark*

'With vivid detail, Dr Robinson has synthesized volumes of international research on microbes – the end result is a concise, reader-friendly page-turner that will be of interest to a full spectrum of readers, from the general public to clinicians to researchers looking for the next big idea! *Invisible Friends* is just like the microbial world that surrounds us – a companion filled to the brim with marvellous potential!'

—Prof. Susan Prescott, author of *The Secret Life of Your Microbiome*

'Can bacteria really be our friends? Thankfully, yes! *Invisible Friends* provides a fantastic, whimsical, and engaging exploration of our friendship with microbes, a friendship ideally built on caring for each other.'

—Justine Dees, PhD, Founder of Joyful Microbe

'*Invisible Friends* tells a wondrous and vital story, how our very being depends upon the unseen, a reality that can help redefine our broken relationship with nature.'

—Miles Richardson, Professor of Human Factors and Nature Connectedness, University of Derby

'A remarkable book by a writer who really is fascinated about the wonders of the world. This book conveys an important message that our invisible friends rule the world and are vital to the health of our planet, including us humans.'

—Marja Roslund, Environmental Scientist, Natural Resources Institute Finland (Luke)

INVISIBLE FRIENDS

INVISIBLE FRIENDS

How Microbes Shape Our Lives and the World Around Us

JAKE M. ROBINSON

PELAGIC PUBLISHING

First published in 2023 by
Pelagic Publishing
20–22 Wenlock Road
London N1 7GU
www.pelagicpublishing.com

Invisible Friends: How Microbes Shape Our Lives and the World Around Us

All illustrations by the author unless otherwise stated.

A CIP record for this book is available from the British Library

Aspects of Chapter 6 were published in the journal *Science of the Total Environment* in 2020. Aspects of Chapter 7 were published in the journal *Frontiers in Psychology* in 2021. Aspects of Chapter 11 were published in the journal *Challenges* in 2018. Aspects of Chapter 13 were published in the journal *Frontiers in Microbiology* in 2021.

ISBN 978-1-78427-433-7 Hbk
ISBN 978-1-78427-434-4 ePub
ISBN 978-1-78427-435-1 ePDF
ISBN 978-1-78427-442-9 Audio

https://doi.org/10.53061/NZYJ2969

Cover design: Laura Brett

Typeset in Chennai, India by S4Carlisle Publishing Services
Printed in Britain by TJ Books

I affectionately dedicate this book to my wife, Kate.

Contents

INTRODUCTION

A Hidden World

As soon as the deep-scarlet sunrise burst from the horizon, the dew-draped meadow glistened like a sea of diamonds. A light breeze carried the scent of a changing season. It was springtime in 2006. I was camping on the semi-forested rolling hills of the Peak District in England. Prematurely awoken by a fleeting rain shower, sleep inertia came to an abrupt end as a potent earthy aroma wafted into the tent. However, at the time, I was completely unaware that my olfactory system – the part of the body responsible for processing smells – was allowing me to perceive signals from another world hidden within our own. A world that we cannot see with the naked eye, brimful of invisible biodiversity, a bustling microscopic metropolis. Initially, I probably thought, 'ahh, the smell of the countryside' or 'the musky scent of a forest' – but then I asked myself, 'Why does it smell like this? What is responsible?' Much to the annoyance of my parents, I have always been an inquisitive soul, asking questions at every opportunity. I trace some of this back to Mr Birch, a slightly eccentric teacher at my primary school in England.

Mr Birch was keenly interested in geology, and we used to swap rocks – our nerdy take on a swap shop. 'You're so cool,' my sisters would say, sarcastically of course. I remember bringing Mr Birch a small chunk of sedimentary rock I had found in the local nature reserve, and him exchanging it for a gnarly piece of igneous rock, or solid lava. 'Wow, is this from a volcano?' I asked. 'Yes,' Mr Birch responded. To my surprise, he told me he had found this volcanic rock in England. I was suitably confused for a seven-year-old. I had travelled around the British Isles to see all the great Norman and Anglo-Saxon castles – one of my father's hobbies – but never once had I seen a volcano. As far as I was concerned, a volcano was a gigantic rocky cone spewing red-hot lava, consuming everything in its path. Mr Birch explained that the area we today know as England once contained many active volcanoes.[1] The rock he gave me was a fragment of one of these, formed thousands of years ago.

I was amazed and still slightly confused. Mr Birch responded with words to the effect that 'There is more to the world than we initially perceive. And remember, the most important word in the world is… *why?*' From this point on, at every opportunity, I would say to my parents and probably everyone around me, 'That's interesting, but why?'… 'Why is this the case?'… 'Why did that happen?' I now see this as both a curse and a blessing. This constant questioning from an untempered child was undoubtedly annoying for the recipient. However, it opened the door to a world of fascination. What else exists in the world that we cannot see? What happens in the world that we may not initially notice? Does what we perceive of the world match the reality? Is everything we thought we knew about the universe even true? These profound questions would both overwhelm and excite me as a child – and they continue to do so even now.

Let's return to the earthy and evocative scent I noticed in 2006. When I arrived home from the camping trip, I remember searching the internet to try and find out what was responsible for that musky smell particular to the countryside, 'the Earth's perfume'. I couldn't find anything. Rudimentary academic literature search engines existed in 2006, but I certainly was not aware of them back then. However, a couple of years later, I discovered that the earthy odour was called *petrichor*, first named by two geologists in 1964.[2] Petrichor is caused by a potent chemical called *geosmin*.

The human nose is incredibly sensitive to geosmin. We can detect this compound at one hundred parts per trillion. In other words, we can detect geosmin better than sharks can detect blood.[3] Geosmin is produced by a group of soil bacteria called *Streptomyces*, and all the known 550 *Streptomyces* species can produce this chemical. This suggests that geosmin confers a selective advantage on the bacteria; otherwise, it is doubtful that all species would produce it. By 'selective advantage', I mean the production of geosmin must bring an essential benefit to the bacteria, and therefore the bacteria evolved to keep this trait and pass it on to future generations.

Some scientists think that *Streptomyces* produce geosmin because of a mutual relationship between the bacteria and a group of six-legged creatures called springtails. These insect-like creatures evolved at least 400 million years ago and are still around today. To disperse across the landscape, *Streptomyces* produce spores – as a plant produces seeds. But how do these spores spread? Well, *Streptomyces* have an ingenious trick up their sleeves. It turns out that geosmin is attractive to springtails, and the bacteria probably produce it as a signal specifically to lure springtails: 'Hey springtail, come and get me!' Scientists think that springtails gobble up the *Streptomyces* as a nutritious food source, spurred on by the fact that *Streptomyces* also produce antibiotic compounds that can kill off pathogens. This is a mutual relationship. The springtails disperse the *Streptomyces*' spores by consuming and excreting them; thus, the bacteria also benefit.[4] This is much like when a badger consumes elderberries and defecates the seeds, allowing them to disperse and germinate – which is why you often find elder trees with their white florets and deep-purple fruits next to badger setts.

I've often wondered why humans have evolved such an acute sense for geosmin. Could it be that humans also receive a health benefit from the *Streptomyces*, and are therefore attracted to the earthy perfume they produce?

Cue inflammation. We've all seen the effects of inflammation on our bodies. It causes that ensuing pain and swelling after a paper cut or a grazed knee. But it also happens inside our bodies as harmful stimuli such as pollution or pathogens trigger a protective response by the human immune system. It's an entirely natural process and involves shuttling immune cells and chemicals to the site of injury or infection, often leading to heat and swelling. Too little inflammation and the harmful stimuli can destroy the body's tissues. Too much inflammation can damage the cell's DNA, leading to severe health conditions.

A 2018 scientific publication highlighted that persistent inflammation contributes to chronic diseases like diabetes and cancer.[5] This is widely known. However, the paper also pointed out that *Streptomyces* produce numerous anti-inflammatory compounds. Importantly, *Streptomyces* species occur in the human gut microbiome – 'microbiome' meaning the entire collection of microbes and their theatre of activity in a given environment, such as guts, armpits, soil or plants. In their conclusion, the authors argued that *Streptomyces* may have evolved to be friendly with humans, helping to suppress colon cancer. So, if this is proven to be accurate, a soil-dwelling bacterium that produces geosmin could also help prevent human diseases. This is only speculation, but perhaps it goes some way towards explaining why humans have evolved to detect geosmin so acutely. By wandering around natural environments and allowing *Streptomyces* spores to hitch a ride, we could also be their dispersers, just like the springtails – another mutual relationship. For me, this fascinating and potentially important relationship demonstrates why we should always ask *why* – even if it's just to ourselves.

Many fascinating phenomena in our world often go unnoticed. The incredible diversity of the microscopic realm around us holds many secrets. It is a shame that we cannot easily see this invisible, bustling metropolis of biodiversity, because it is always challenging to appreciate what you cannot see. However, technological breakthroughs and the rapidly plummeting costs of machines that decipher the building blocks of microbial life (DNA) allow us to understand the dynamics and interactions of this hidden world.

Rapid technological advances have opened the door to many new scientific endeavours, from sequencing the human genome to editing genes for therapeutic purposes. Microbial ecology, now a booming field of research, has capitalised on these breakthroughs. We now know the air we breathe is thronging with microscopic life-forms: moss spores and plant pollen, dense clouds of bacteria, archaea, tiny fungi and algae, along with protozoans and vast quantities of viruses, each communicating, interacting, sharing and competing all around us. There may even be a few microscopic

Moss piglets (tardigrades) often live among moss and lichens.

moss-dwelling animals called tardigrades, also known as water bears or moss piglets because of their mammal-like appearance – under the microscope, at least. A single gram of moss on the forest floor may contain over 100,000 of these tiny animals, and because they are exceedingly light, they are easily swept up by gusts of wind, making them airborne.

Researchers can now analyse vast quantities of microbial samples, providing novel insights into the complex but unseen communities around us. I have done this myself as a researcher, in the course of completing a PhD in microbial ecology.

In this book, I aim to reveal the weird and wonderful roles microbes play in shaping our health and behaviour, and indeed in the wider world around us. Highlighting the etymology of the mainly Greek and Latin scientific names is another thread running through the book. The roots of words provide associations that allow us to remember the complex names of species or concepts, but they also indicate their function and sometimes environment. For instance, *Thermus aquaticus* is a species of bacterium that grows in hot (*Thermus*, from the Ancient Greek *thermós*, 'hot') springs (*aquaticus*, from a Latin word relating to water). It was first 'discovered' in Yellowstone National Park in the United States back in 1969.

Incidentally, *T. aquaticus* played a pivotal role in advancing DNA technology and thus the field of microbial ecology. The bacterium is the source of a heat-resistant enzyme that can withstand the high temperatures in the 'PCR' (short for polymerase chain reaction) process. PCR is a method of amplifying small samples of DNA so that scientists can study them in detail. It can target a specific gene conserved in all bacteria, allowing one to inspect only the bacterial members of a sample (which is often teeming with DNA from other microbes, plants and animals). It can 'pull a needle out of a haystack', enabling a rare component of a large and messy mixture to be identified. Indeed, it's the same tool widely used to test for COVID-19 infections. The chemicals used in a PCR reaction hook onto a segment of the virus's genetic material so that it can be amplified. Therefore, in a sense, we have a hot spring-loving bacterium to thank for our ability to fight against COVID-19 and other maladies.

Historically, our general perception of microbes has been negative due to the relatively few invisible foes that cause diseases. In this book, my intention is not to play down the severity of pathogens. Instead, I intend to stimulate a more balanced view of microscopic life-forms and showcase fascinating stories about their underappreciated and often beneficial roles in all aspects of our lives.

Let's cast our minds back to the dawn of germ theory, which Hungarian physician Ignaz Semmelweis anticipated and French chemist and microbiologist Louis Pasteur consolidated in the mid- to late nineteenth century. It was a remarkable development in human thinking when scientists understood that microorganisms were responsible for many human maladies. Before this, our understanding of human diseases was limited. One of the predominant theories in the Western world was the miasma theory.[6] This held that diseases such as cholera or plague were caused by a noxious form of 'bad air' emanating from rotting materials. Germ theory, on the other hand, states that some microbes can cause diseases by invading the hosts (whether humans, other visible animals and plants, or even other microbes) and causing physiological harm. People often refer to microbes as 'germs' (from the Latin *germen* for 'seed', 'bud' or 'sprout'). Presumably this is due to their budding-like lifecycle and seed-like appearance. Knowledge of pathogenic microbes has undoubtedly saved millions of lives since germ theory was first posited. However, knowing that some microbial species – though in truth it is far fewer than one in ten thousand – cause human diseases has led many people to fear and loathe all microbes.[7] This 'germophobia' has likely been compounded by decades of relentless advertising campaigns, such as those selling household detergents or that instil fear of nature and dirt. These campaigns have created a negative perception of all microbes.

Often, the result of this perception is the avoidance of natural environments and their dirt, the mass sterilisation of surfaces with detergents, and reduced human exposure to biodiversity – the variety of life around us, including the invisible kind. It turns out that this avoidance of biodiversity could be contributing not only to a loss of appreciation for the vital, invisible universe around us, but also to an explosion in human immune-related disorders.[8] So, wiping out all the microbes in our lives with a view to preventing diseases could be having the reverse effect. It is important to note that targeted hygiene remains essential around food, sinks and toilets. However, attempting the total elimination of dirt from our lives is where the danger lies.

The truth is that relatively few microbes cause human diseases, and many others benefit us. Bacteria and bacteria-like organisms called archaea, along with algae, fungi and tiny animal-like critters called protozoans – and even some viruses – all play vital roles in our ecosystems. They are the glue that holds it all together – our invisible friends. Indeed, microbes play essential roles in plant health and communication, animal health, nutrient cycling and climate regulation, among many other ecological processes. Without microbes, our food systems would collapse. Without microbes, our societies would crumble, and our bodies would not function in the long term. As microbiologists Gilbert and Neufeld said in a paper published in 2014, 'if we include mitochondria and chloroplasts (the energy-producing cells in animals and plants) as bacteria, as we should, then the impact [of their absence] would be immediate – most [creatures] would be dead in a minute'.[9]

I want to challenge the prevailing negative perception of microbes and take you in a different direction, shining a light on all the fascinating roles that microbes play in our daily lives and discussing their relationships with all other life on Earth. In what follows, we will hear from world-leading experts in microbial ecology, neuroscience, restoration ecology and immunology, and I even visit a regenerative agriculture farm. I discuss some of the risks to our relationship with our invisible friends, such as antimicrobial resistance, the biodiversity crisis and the rise in germophobia mentioned earlier, in addition to framing microbes as a facet of social equity (Chapter 4). Microbes are essential features of our ecosystems, health, social structures, behaviour, food systems and cultures. They are infinitely small but do infinitely great things.

Stories of interconnectedness weave their way into each chapter, and there are occasional philosophical musings around our affinity with the natural world. I believe we need to redefine our relationship with nature culturally, socially, psychologically and emotionally – particularly in Western societies. Still, there's scope for all people in all communities to redefine their relationship with nature *microbiologically*, through knowledge of the unseen cosmos outside and in. I hope you enjoy reading about our invisible friends.

For those who would like to know more about microbes before reading this book, I've included an appendix called 'Microbes 101'. If you're new to the world of microbes, this should help with some of the terminology and references used.

CHAPTER 1

The Microbiome

'The role of the infinitely small in nature is infinitely great.'
—Louis Pasteur

I'm sitting on a rickety old camping chair in a pine forest, at the top of a steep hill. The forest receives few visitors. It has a calming aura, helped by the trickling sound of a boulder-hugged stream. This is where I come to clear my mind, and sometimes to work. It could be to write. It could be to zone out and run code on my laptop. It could be to draw some inspiration for new research or simply to reflect, find solace and, as nineteenth-century naturalist John Burroughs said, 'to have my senses put in order'. The forest is a beacon of serenity.

The ground beneath my feet is carpeted with the creeping shamrock-shaped leaves of the edible wood sorrel. The air is rich with buzzing hoverflies, shield bugs and ladybirds. My eyes drift left, right, up and down across the trees' diverse contours, textures and pleasing fractal patterns. I find myself considering the rich bounty of ecological niches and elegant adaptations surrounding me. And how each individual from the consortia of plants and animals I see with the naked eye is a diverse conglomerate of many organisms, the vast majority of which I cannot perceive. The trees are host to trillions of microbes. The trees need the microbes for development, communication and, ultimately, their survival. The mosses that creep across the boulders are also home to trillions of microbes, as are the wood sorrel and the hoverflies, and my solitary self. But this means I am in fact anything but alone. My body is a hive of activity, a bustling jungle full of life. I sit here emitting my personal signature in the form of a microbial cloud, and I bathe in the microbial clouds of the plants and animals around me. Our microbiomes are in constant flux, and constant communication.

Before we go any further, I should define some terms. For instance, what is a microbe and a microbiome? A microbe, also known as a microorganism, is a microscopic organism that can either be single-celled (unicellular) or multicellular. The word comes from the Greek *mikrós* ('small') and *bíos* ('life'). Microbes include both prokaryotes (pronounced 'pro-carry-ohts') and eukaryotes (pronounced 'you-carry-ohts').

The *prokaryotes* are single-celled creatures that lack both a membrane-bound nucleus (where DNA is stored) and organelles – which literally means 'tiny organs'. Prokaryotes include bacteria and similar-looking microbes called archaea. *Eukaryotes*, on the other hand, do have a membrane-bound nucleus and organelles. Microscopic eukaryotes include fungi, algae and protozoans – tiny animal-like creatures.

Lastly, and by no means least, there are the viruses. These are neither prokaryotic nor eukaryotic. Viruses can only replicate within the cells of a host creature. As such, scientists often give them the unflattering description 'obligate parasite'. Essentially, microbes include any organism you would need a microscope to see, along with viruses, which most people consider to be non-living entities.

Microbes are also incredibly diverse and incomprehensively abundant. For example, current estimates suggest 10^{12} microbial species exist on Earth; that's one trillion different species.[1] For those who enjoy snappy analogies, there are thought to be ten times as many microbial species on our planet as there are stars in the Milky Way.

Biodiversity – or the variety of life on Earth – is so much more than meets the eye. It would be easy to inadvertently perceive biodiversity, from which we acquire a rich bounty of provisions and aesthetic values, as simply the trees, flowers, insects, birds, amphibians, reptiles, mammals and other wondrous visible life-forms that inhabit the planet. After all, these are the organisms that we can see, as well as hear and sometimes feel. There are an estimated eight million species of meso- and macroscopic (visible) animals and plants on the planet. However, dig a little deeper, and we find this figure is dwarfed 125,000 times over by the number of different microbial species. And speaking of digging deeper, if I were to move the wood sorrel beneath my feet to one side and plunge a teaspoon into the soil, I would likely return with between 10,000 and 50,000 different microbial species, or one to seven billion individuals, on the spoon. To cite a quote often attributed to Leonardo da Vinci (1452–1519), 'we know more about the movement of celestial bodies than about the soil underfoot'. This could still be true today, although we are at last catching up, thanks to rapid advances in technology.

Microbes form complex and dynamic communities, much like the so-called 'higher organisms' (a rather grandiose title often given to larger and visible animals and plants), and they inhabit all the world's ecosystems. Some microbes are uniquely adapted to extreme environments where others would simply fail to survive. For example, if I took a bacterium from the mossy boulders next to the woodland stream and placed it in a hot spring, it would be unlikely to survive, and vice versa – a specialist hot-spring microbe would not fare well in the chilly temperate forest moss. Still, many other microbes are 'generalists' and can adapt to a wide range of environmental conditions. In addition to the tremendous number of microbial species and the variety of their ecological niches, a diverse range of shapes and sizes exist. Some are tubular, spherical, crowned or filamentous, and others form long chains. Some are rod-shaped, and others are 'icosahedral' with 20 triangular faces. Bacteriophages, or viruses that prey on bacteria, are often depicted as spider-like spaceships complete with landing gear. This is quite an accurate description, as shown in the sketch below.

The smallest known microbe on the planet is a virus, and it is approximately 20 nanometres (nm) in diameter. Discovered in 1965 by R.W. Atchison,[2] we call this an adeno-associated virus or AAV. The largest known microbe is a bacterium called *Thiomargarita namibiensis*. It's a spherical bacterium around 0.75 mm in

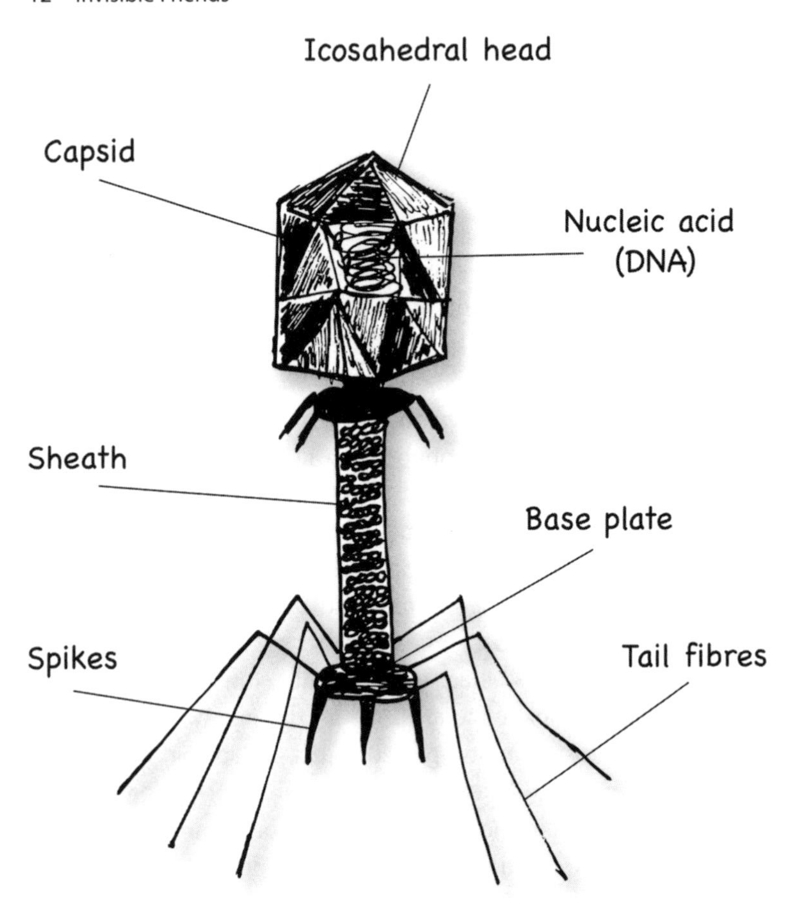

The alien spider-like bacteriophage, or 'phage' for short. This is a sketch of the T4 phage that infects *E. coli* bacteria.

diameter and is found in ocean sediments off the coast of Namibia. This one is easy to notice because it shines like a pearl, which gives its colloquial name, the 'sulfur pearl of Namibia'.[3]

The size difference between the virus and the bacterium might not seem so vast. However, 20 nm is 0.00002 mm, and there are 750,000 nm in 0.75 mm. Therefore, the bacterium is 37,500 times bigger than the virus. If the virus sat down on the bacterium, this would be the equivalent of a human lying down inside the boundary of the M25 motorway in London, which has a diameter of 75 km. It's also the equivalent of placing a poppy seed on top of a blue whale – the largest animal on the planet. I had to re-run this calculation several times before I believed it myself. But we shouldn't let the diminutive size of viruses fool us – they really do pack a punch. In fact, every 48 hours, phage viruses kill half of all bacteria on the planet.[4] They play vital roles in maintaining the biological equilibrium of ecosystems – including that in our bodies. Indeed, a healthy person is home to a hundred times more viruses than there are individual trees on the planet – a hundred global forests of phages inside you![5] Many of your 380 trillion viruses play a role in controlling bacterial populations. Hence, it's likely that we need them, just as they need us; sweet reciprocity between our bodies and our residents, unbeknown to our minds. And there's another – perhaps surprising – reason we should tip our hats to viruses: their role in providing the air we breathe. Take a deep breath. Each time we draw air into our lungs, around 21% is oxygen. Photosynthetic organisms have been releasing this oxygen into the atmosphere for over 2 billion years. Some of them are plants (in the last 500 million years), but many are microbes – particularly in the ocean. While the ratios have changed over time, photosynthetic cyanobacteria in the ocean generate at least half of the oxygen in the atmosphere.[6] But where do viruses enter the equation? Scientists think that the photosynthetic machinery in the cyanobacteria is of viral origin.[7] Because viruses can transfer genes between different organisms like prolific market traders, they may well have played a crucial role in cyanobacteria evolution. This includes their ability to photosynthesise and thus produce oxygen. Moreover, scientists believe some phage viruses insert photosynthesis genes to keep the host cyanobacteria on 'life support' during infection.

Draw in another deep lungful of oxygen and reflect: a diverse team of oceanic bacteria and viruses made this possible.

Microbiome

What is a microbiome? You might describe a microbiome as 'the entire collection of microorganisms in a particular environment'. This is a simplified definition. It sounds relatively straightforward. However, molecular biologists might define it more precisely as 'the collection of microbial genetic material in a given environment'. Others will include environmental conditions and different aspects of the ecological theatre in which the microbes operate.[8] For clarity, in this book, I will be using a definition that encompasses this broader idea – so, all the microorganisms in a given environment, and their entire ecological theatre of activity. The word microbiome comes from the Ancient Greek *mikrós* ('small'), *bíos* ('life') and *ōma*, meaning a 'complete or collective body of'.

We can trace theories about the interactions between different microbes back to the late nineteenth century, and perhaps the first *microbial ecologist* was Ukrainian Sergei Winogradsky. In 1888, he recognised and studied the interconnections between aerobic (oxygen-requiring) and anaerobic (requiring an absence of oxygen) bacteria.

Despite steady advances in microbiology during the twentieth century, it was only in 2007 that microbiome studies gained rapid momentum with the launch of the Human Microbiome Project. This US government-funded initiative aimed to conduct the most comprehensive assessment of the microbial communities living in and on the human – a huge ecological survey on a microscopic scale. A subsequent initiative, the Earth Microbiome Project (2010), did the same for global ecosystems. These projects have inspired many of the insights in this book.

I continue to sit in the forest with a vivid picture in my mind. I visualise the ground beneath my camping chair humming with activity. With every breath I take, I picture a constant flux of life-forms being exchanged between myself and my surroundings. All these life-forms – the bacteria, archaea, fungi, viruses, algae and protozoa – often live together in complex dynamic communities, like the microbiomes of the pine forest I am sitting in. I look at the trees next to me. The aboveground part of the trees is referred to as the phyllosphere (pronounced 'fill-lo-sphere'), and it has a distinct microbiome. I look down to the soil beneath my chair and see the area surrounding the roots; this is known as the rhizosphere (pronounced 'rye-zo-sphere'), which also has a distinct microbiome. The microbiome of the air I breathe, meanwhile, is known as the aerobiome. Much like the habitats around us, our bodies also harbour distinct microbiomes. Indeed, we could view ourselves as *walking ecosystems*.

Humans as walking ecosystems

Like the trees in the pine forest, we too have our own microbiomes. Each of us has several distinct microbial communities living in and on our bodies. Microbiome researchers say that the microbial genes in our bodies outnumber human genes by 150 to 1, and if we could line the microbes from a single person's gut up end to end, they would be able to circle the Earth two and a half times. As Walt Whitman said, 'I contain multitudes' – which by no coincidence is also the name of a fantastic book on the human microbiome by the science journalist Ed Yong.[9]

A complex and dynamic conglomeration of molecules, cells, tissues and organs clinging on to a gangly, mineralised frame and working harmoniously like a quintessential symphony orchestra. I am, of course, referring to the human body. Add a potent sprinkling of consciousness to the mix, and you have a genuinely momentous system capable of unimaginable feats. Yet, as complex as human beings are, we are descended from the single-celled archaea. Jump into a time machine and travel back far enough – some 3.5 billion years – and you will find your ancestors are like those diminutive but extraordinarily resilient germs we often discover in so-called extreme environments today, such as hot springs and salt lakes.

The next revelation is perhaps equally peculiar. Within each of our cells are hard-working organelles called *mitochondria*. These bean-shaped intracellular structures are responsible for producing a chemical known as ATP (adenosine triphosphate) through a tongue-twister of a process called oxidative phosphorylation. These tiny organelles produce over 90% of our cellular energy. There's no wonder they're often called the 'powerhouses' of our cells. Yet we think mitochondria also evolved from bacterial ancestors. It may seem surprising that the tiny organs that provide energy to our cells and are essential to respiration are *microbial* in origin. Yet scientists increasingly accept that this is indeed the case. Many scientists consider an internal symbiosis (where two different organisms live in close physical association) to be the leading theory for how mitochondria evolved; the mechanism for this phenomenon is called *endosymbiosis*.[10] One theory states that a bacterium was phagocytosed (aka 'gobbled up') by a eukaryotic organism. This eventually led to a remarkable symbiotic relationship whereby one organism lives inside the other.

By now, you may have noticed a theme of interconnectivity weaving its way through the narrative. All things connect, whether socially, biologically or evolutionarily, and microbes are the glue. As humans, we are intertwined with microbes, and throughout evolutionary history, they have shaped our bodies and minds. They were here billions of years before us, curating our DNA, our cells and the 'metaorganism' we embody today. Despite our egos, which may tell us otherwise, it seems *they* are the dominant life-forms on the planet. Our fleeting existence will not ensure our resident microbes live for an eternity, but there is no doubt they will find another home once we are gone.

Now that we understand the importance of microorganisms, we have an opportunity to redefine our relationship with the natural world. Indeed, there is a whole cosmos of invisible biodiversity out there, and with the help of next-generation DNA sequencing, we can look much closer to home to find a diverse array of species – within and upon the human body. In recent years, researchers have collected samples from various locations in living human bodies to analyse the content and identify microbial DNA signals. The results: around 150 microbial species on our hands, 700 in our mouths, and 1,000 in our guts.[11]

We each emit somewhere in the region of a million biological particles (including microbes and 'human' cells) per hour from our breath, skin and hair. The presence of just one person in a room can add 37 million bacteria to the air every hour from emission and displacement. We can view ourselves as dynamic ecosystems – a human host plus trillions of microbial symbionts (organisms that live in close association with others) openly interacting with the environment via complex biological exchanges. Indeed, we are deeply embedded within the natural world – both physically and psychologically – and we are highly dependent on biodiversity at multiple scales. In just a cubic centimetre of our walking ecosystem (our bodies), millions of collaborating and competing microbes contribute to nutrient cycles and energy flows without us even noticing.

As I sit on my camping chair, admiring the life around me, I feel like I've metamorphosed from an object into a group of subjects. Like a psychedelic trip, albeit fleeting, this process of introspection has momentarily altered the contours of my consciousness.

A hoverfly lands on my arm. Moments before, it was perched on a creaking Scots pine tree. Before that, on the boulders that hug the stream. I've become acutely aware that the hoverfly delivers and collects a new group of microbes with each stepping stone it lands on in this forest. Microbial life-forms even adhere to the hoverfly's body as it sails through the air. And now it has landed on my arm.

Touch-down. This ultra-light organic connection allows a flurry of microbes already on my arm to join the hoverfly's personal ecosystem, and for the hoverfly's microbes to join mine. Yet the microbes joining my body didn't come from just the hoverfly, but also from the pine tree, the mossy boulders and the air: a phenomenal multi-species correspondence, a biological internet. The term *constant flux* springs to mind over and over. We are all in this together. We are all connected through our invisible friends.

CHAPTER 2

Rekindling Old Friendships in New Landscapes

'We mostly don't get sick. Most often, bacteria are keeping us well.'
—Bonnie Bassler

Once upon a time, most multicellular life-forms were slimy or spongey. Then around 500 million years ago, animals evolved to have a backbone. At the time, microbial communities living in and on animals became more numerous and complex. The first vertebrates (animals with a backbone) were jawless fish. They lived in the primordial ocean abyss. It was another 300 million years or so before mammals appeared on land.

Scurrying around the dinosaurs of the time were fleet-footed mouse-like creatures and slightly larger animals resembling modern-day badgers.[1] Indeed, the first land mammals were burrowing creatures, much like today's voles, moles and prairie dogs. Our ancestors were literally coated in soil; they breathed in and even consumed it. Within the soil lived trillions of microbial denizens, meaning that mammals co-evolved in close-knit relationships with soil microbes for millions of years. As we will see later in this chapter, the co-evolution of vertebrates and microbes helped to shape the mammalian immune system.

As humans, we have since partially removed ourselves from our ancestral natural environments, particularly in so-called high-income nations. We are simultaneously the most connected and the least connected we have ever been. We are connected to many others via the internet (a shallow synthetic connection), but the deep organic bonds that connect us to the rest of nature are breaking. Many of us now reside in brick, metal and plasterboard cubes. Indeed, many of us live in these cubes stacked on top of one another, hundreds of metres high in the polluted ravines of the concrete jungle. This has likely contributed to a rise in immune-related diseases and has deepened the gulf between humans and the rest of the natural world. To reverse these immune disorder trends, we need to rekindle our old friendships with microbes, albeit in new landscapes.

I first met Professor Graham Rook at the beginning of my PhD, travelling to London on the train from Sheffield to interview him in his home for a podcast called *A Dose of Nature*. I decided to interview Graham again for this chapter, but this time, it had to be a video call due to COVID-19-related restrictions.

Graham grew up in Cambridge – a place that some would consider one of the hearts of academia. His father, Arthur Rook, was known as *the* dermatologist of his time, and he wrote a revolutionary textbook on dermatology (the science and medical study of the skin).[2] Arthur had a major influence on Graham's developing passion for science. Graham had always experimented with research in his bedroom as a child. Indeed, his bedroom doubled as a laboratory of sorts. He even dabbled with electrophoresis on the floor. This is a process whereby DNA (the building blocks of life) and other molecules are separated based on their size and electrical charge.[3] Lacking a fully equipped home laboratory, Graham improvised; he created an electrophoresis apparatus using an electric heater as a resistor to alter the voltage. He also used a rectifier – a strange and now-archaic contraption that only allows electrical currents to flow in one direction. Like many who end up in the field of biology, Graham also played around with microscopes, looking at creepy-crawlies from water samples. 'Isn't pond water the beginning and the end of everything?' he joked.

Graham studied basic sciences and immunology at Cambridge University. After this, he went to study medicine at St Thomas's Hospital in London. At the time, immunology (the science and study of the immune system) was the newest and one of the most exciting subjects available. Therefore, despite Graham being hugely influential in the field of microbiology, he is actually an immunologist. Following his training, he worked at the Middlesex Hospital in London for a while, which had a world-leading immunology department. Despite having just got started, the immunology field was expanding at a dramatic pace. Graham recalls often leaving immunology seminars shaking with excitement.

For many years, he worked on tuberculosis, also known as TB. This is a highly infectious and deadly bacterial disease that typically affects the lungs.[4] The bacterium responsible for the disease is called *Mycobacterium tuberculosis*. A cousin of this bacterium played an altogether different but prominent role in Graham's illustrious career, which I'll return to later. Incidentally, some clues about the nature of the organism that causes TB can be found in its scientific name – for the Latin *tuber* (meaning 'a

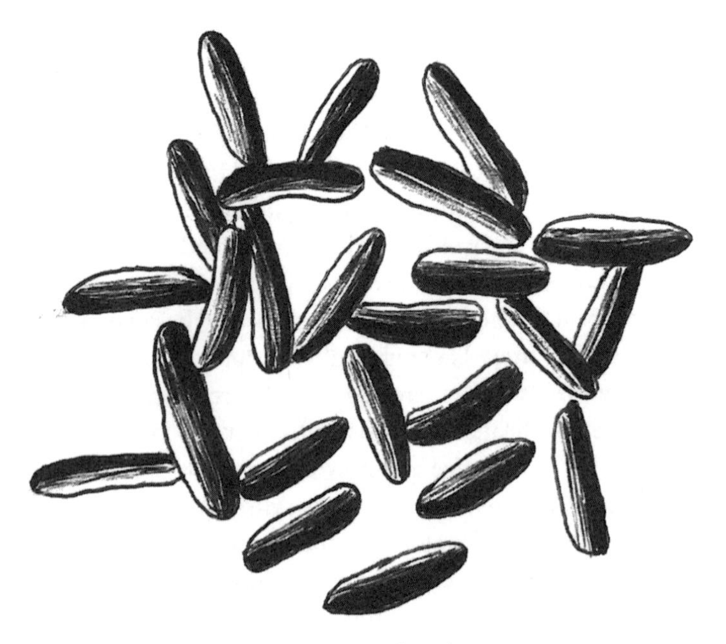

Mycobacterium tuberculosis.

hump or swelling') describes its shape and the Greek prefix *myco* ('fungus') alludes to the way the bacterium grows in a fungus-like manner.

Graham's team would carry out experiments to help immune cells called *macrophages* slow down the growth of TB. Macrophages are widespread in the body, occurring in almost all tissues. The word macrophage comes from the Greek *makrós* ('large') and *phagein* ('to eat'). These etymological roots describe the ability of macrophages to engulf and digest anything apart from normal and healthy body cells. In other words, macrophages are the primary line of defence against invading pathogens. However, Graham light-heartedly admitted that 'in the decades that we worked on TB, neither we nor anyone else working in the field got anywhere close to improving the treatment of the disease!'

Graham eventually worked with other species of *Mycobacterium*, particularly those found in soils. He and his colleagues made some fascinating discoveries about *Mycobacterium*. For instance, they could tell where someone was born, whether in Kent, UK or Kampala, Uganda, simply by doing a series of skin tests and assessing the person's immune response. If the person already had *Mycobacterium* antibodies (immune cells that bind to toxins), it was highly likely that they had already encountered the *Mycobacterium* toxin. Importantly, this toxin is specific to the location or environment it evolved in, hence the ability to predict the person's homeland.

Imagine for instance when you give your pet cat a treat, and you use a particular tone of voice – probably a high-pitched one – to say something like 'Do you want a treat?' Your cat has already been exposed to this signal and responds by getting excited and meowing fervently. This is like the immune system responding to a toxin it has previously encountered. But if you said 'Do you want a treat' in a slow, low-pitched and monotonous voice, the cat hasn't been exposed to this signal, and so it probably won't respond in the same manner.

Later in his career, Graham became heavily involved in neuroscience, after working with Dr Chris Lowry from the United States. Chris and Graham experimented by immunising mice with another species called *Mycobacterium vaccae*. They found that, after inoculation, the mice's neurons lit up like fireworks. And when they tried to stress the mice, it didn't happen – the *M. vaccae* bacterium regulated the stress response. This was a remarkable discovery.

And it didn't end there. They also found that the bacterium had anti-allergic properties.[5] The microbe, which came from the soil, had all kinds of immune-system-regulating and stress-reducing effects. This was an enlightening moment for Graham. He had always been interested in evolution, and he began looking at this process from an evolutionary perspective.

In the late 1980s, Graham read a paper on the *hygiene hypothesis*.[6] This is still a widely accepted concept, though, as we will see, an update may be needed. The article discussed how children around the age of eleven were less likely to have allergies if they had siblings. The theory that emerged was that we must need exposure to the common infections of childhood to build a robust immune system and prevent allergies. However, this theory didn't chime strongly with Graham. He thought there was no way we could have evolved dependence on the common infections of childhood. Indeed, he knew of one estimate suggesting that measles (a highly infectious viral disease) didn't even infect humans until the twelfth century. As he tells me, 'it is more likely that it was sometime during the Roman Empire when there were sufficiently large populations for these diseases to become endemic.'

Graham explains that we now refer to the common infections of childhood as 'crowd infections'. COVID-19 is another crowd infection that has recently taken advantage of our vast population and crowded lifestyles, although it's still far less contagious than measles – see the diagram below.

For much of our evolutionary history, crowd infections didn't exist – simply because there were no large crowds. For thousands of years, most humans lived in small bands, hunting and gathering food, and moving from one location to another. Therefore, Graham maintains, we could not have evolved dependency upon these crowd infections to develop robust immunity. To Graham, this meant the crowd infection theory underpinning the hygiene hypothesis didn't make sense. He thought that there must be a

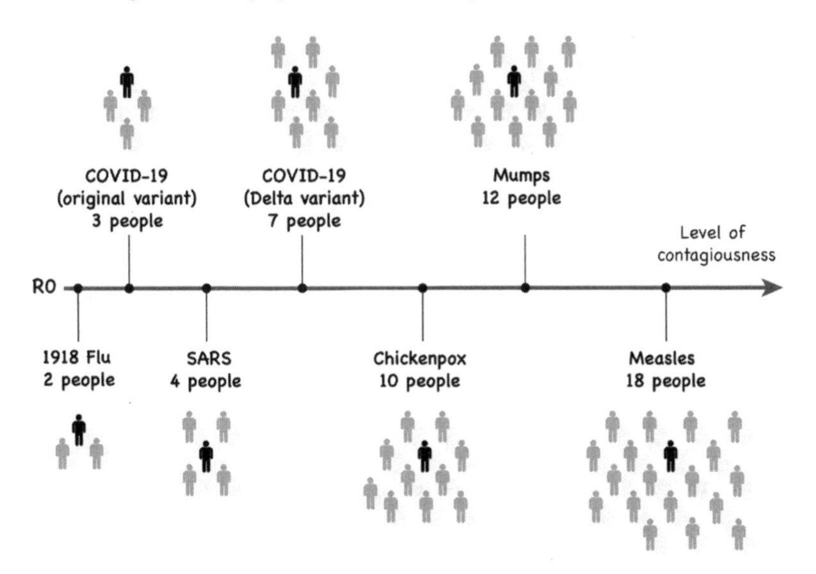

Examples of crowd infections and their relative contagiousness – indicated by the R0 (R-zero) or 'reproduction number'.

different, albeit a microbe-related, explanation for this rise in immune disorders.

Several studies have implied that having common childhood infections, if anything, might make things worse – you're more likely to have allergic problems.[7] Theories surrounding the hygiene hypothesis later switched to the idea that household cleaning agents were wiping out the microbes needed to stimulate proper immune regulation, thus causing allergies.[8] However, Graham points out that researchers have since demonstrated that cleaning products such as bleach damage our body's surfaces. This includes our skin and respiratory lining. And importantly, our immune system reacts to this by mounting a 'kick it out quickly' response to whatever it can detect. Because our immune system can't mount an appropriate response to bleach itself, it will direct its immune response to whatever is present in the environment and on the skin – such as allergens on the hair of the neighbour's cat or the dust mites in the carpet. This could explain the observed correlations between cleaning agents and allergies.

The old friends hypothesis

Enter evolution. Graham tells me that this represents the other side of the coin. 'The idea that microbes in a modern home are somehow essential for our health is wonderfully absurd when you think about it from an evolutionary perspective.' We evolved to have an intimate relationship with the microbes from natural environments, not the microbes in contemporary homes. Most houses are no longer reconfigurations of the natural environment; that is, they're not made of wood, straw, mud and clay. In Western nations, if we have exposed wooden beams in the house, they are probably treated with toxic 'biocide' chemicals that purge the microbes in the wood. Simply put, we have not evolved with the microbes of the typical modern home. This causes all kinds of problems for our immune system. Indeed, the microbes in our houses are only beneficial to the extent that they resemble those from the natural environment. Therefore, it is important to keep the sinks, toilets and food preparation areas clean to remove the 'nasties', but also to ensure the interior of our houses resemble the natural environment in order to host the 'old friends'.[9]

But how do we ensure today's homes resemble the natural environment? Some ideas include having plenty of house plants, keeping pets from a young age and regularly visiting natural environments to (a) receive an immediate dose of naturally diverse microbes and (b) bring them back to the home environment. Growing your own food will also bring diverse microbes into the home and your diet. Living in more natural surroundings is also likely to be important – but this is simply not possible for many people. The fact that the environments surrounding millions of peoples' homes are unnatural (i.e. they don't support plants and animals) underscores an important social injustice. I'll return to this in Chapter 4. Some scientists, including myself, consider restoring natural environments in our towns and cities to be a public health intervention for this very reason – restoring our relationship with our co-evolved microbial old friends.

Using this evolutionary thinking, Graham proposed a replacement to the hygiene hypothesis in the early 2000s.[10] The replacement hypothesis postulated that it is the removal of natural biodiversity from our lives and the lack of interaction between ourselves and the microbes we co-evolved with that causes immune-system issues and inflammatory diseases like allergies. Graham called this the *old friends hypothesis* – see below for a comparison of this and the earlier hygiene theory.

Graham gave the hypothesis the name 'old friends' because of the intimate bonds we've established with certain kinds of microbes over our evolutionary history. Many microbes are 'friends' because they educate and regulate our immune system. We depend on them to fulfil a plethora of roles in our bodies – our walking ecosystems – so that we can function and ultimately survive. In return, we provide them with a home and nutrients. It's a mutualistic relationship – a form of symbiosis.

Graham points out that a recent study demonstrated a strong relationship between the number of cats and dogs in the home and a decreased likelihood of developing allergic disorders.[11] Researchers think this is because our furry companions bring the natural environment into the house. Indeed, when they do this, they likely carry a diverse bunch of microbial hitchhikers that help regulate our immune system.

Hygiene hypothesis	**Old friends (or biodiversity) hypothesis**
Reduced exposure to 'crowd infection' pathogens as a result of being too 'hygienic' could explain the rise in twentieth-century immune disorders	Reduced exposure to diverse microbial communities from natural environments could help explain the rise in twentieth-century immune disorders
Graham argues this would require an evolutionary dependency on crowd infections, but that we couldn't have evolved with these infections because 'crowds' are a recent development	Reduced exposure to 'old friends' (microbes we've co-evolved with) from natural environments could help explain the rise in twentieth-century immune disorders
Studies show a correlation between the use of cleaning agents and allergies	Studies show links between reduced biodiversity and increased immune disorders

The hygiene vs the old friends hypothesis.

Environmental microbes, such as those in the soil, air and plants, play critical roles in training our immune systems. This helps to provide a robust defence against invading pathogens. Graham uses a wonderful analogy for this training process. He likens the human immune system at birth to a computer. Like a computer, our immune system has hardware – the cellular and organ structures. It also has software – the genetic material encoding a given function. But the one crucial thing the immune system lacks at birth is a set of data. Just like computational models (e.g. when computers are used to study complex systems), our immune systems require training data.

There's a famous saying in computational science, particularly when referring to models: 'garbage in, garbage out'. This term expresses the notion that incorrect or low-quality input (garbage) will invariably produce a faulty output (garbage). It is much the same with the immune system. Suppose it doesn't have exposure to an appropriate level and enough diversity of data, particularly during its critical developmental years in early life. In this case, it will likely be a faulty immune system in the long term. The primary source of data I am referring to is the vast array of microbes from our natural environments.

A diverse suite of data in the form of microbes must be supplied to the human immune system at birth and throughout the early years of life. These microbes come from the environment and other human beings. If they are not adequately supplied, the mechanisms that regulate the immune system can ultimately falter and fail. The immune system will continue to attack harmful pathogens when this happens, although it may not be as efficient.[12] However, it will also attack innocuous agents such as pollen, food and animal allergens, and house dust, which manifest as allergic disorders. Arguably, an even worse scenario may occur if the immune system is not adequately primed and regulated: it can attack the body's own cells. When this occurs, we call it *autoimmunity*.

Autoimmune disorders such as rheumatoid arthritis, inflammatory bowel disease and coeliac disease can be severe, and evidence suggests they have increased in prevalence in recent decades. Coeliac disease occurs when the body's immune system overreacts to gluten, which is found in many foods including pasta and bread. We now know it can strike at any age. In fact, my mother was recently diagnosed with coeliac disease having had no symptoms earlier in life. It's genetically linked; if a parent has it, the child has a one in ten chance of developing it. Studies have shown higher levels of protective bacteria such as *Bifidobacteria* and *Lactobacilli* in the guts of people without coeliac disease, suggesting a potential relationship between microbes and the disease. Researchers think that stress, antibiotics, poor diet and certain environments might be damaging our gut microbiomes, leading to poor immunity.

Perhaps unsurprisingly, given our growing knowledge, the rise in autoimmune diseases and other chronic illnesses coincides with the global trend of biodiversity loss (the extinction of plants, animals and microbes). This loss of biodiversity, along with our weakening connection to the natural world, means we are probably losing touch with our microbial 'old friends'.

It is essential to say here that, unlike invertebrates (such as insects and worms) which have *innate* immunity, the human immune system comprises two critical subsystems. One is the innate immune system. The other is the *adaptive* immune system, which I'll come back to later.

These immune-system names and processes can be a bit overwhelming, but bear with me; an analogy will be along in a moment.

The innate immune system

Our innate immune system is fast-acting and is known as 'non-specific'. This means it will attack all incoming substances without appropriate regulation. Our microbial 'old friends' from the environment help provide this regulatory role. They can stimulate chemicals that prevent our bodies from attacking our human cells or those ordinarily innocuous substances like pollen and dust. When we have appropriate exposure to these microbes, our innate immune system recognises them because they are old buddies, evolutionarily speaking. Instead of initiating an aggressive immune response, they ensure our immune system retains the ability to control special cells called 'Tregs', which provide an anti-inflammatory role. If Tregs are missing, the immune system can overreact and we ultimately become ill – it's a fine balancing act.

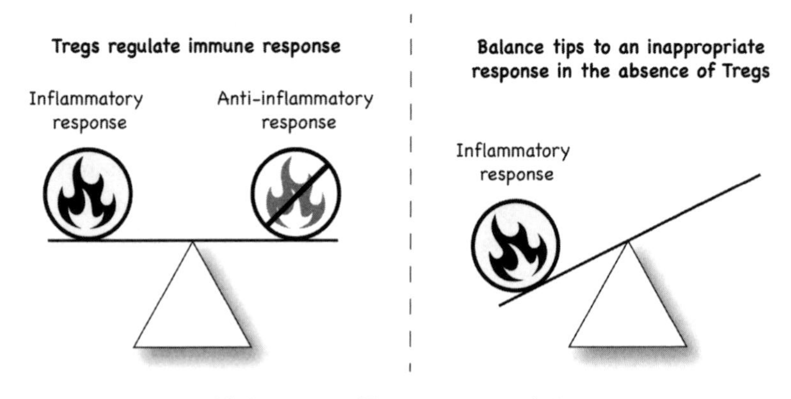

A balancing act: Treg immune regulation.

It is worth remembering here that inflammation is a normal part of a healthy immune response to trap pathogens or to start the tissue-healing process. The result can be pain, swelling, bruising or redness. Too much inflammation is where the problem lies.

In summary, it is thought that if we don't have appropriate exposure to microbial old friends from natural environments, which in turn promote Tregs, our immune system will attack the things it's not supposed to, like dust, pollen, nuts and pet allergens. This is what defines an allergy – an inappropriate immune response to seemingly innocuous particles.

Imagine that you are walking into a high-security building with those walk-through gates that beep loudly and light up whenever metal is detected. And now, imagine the gates could not tell when a person didn't have metal on them and that they interpreted their harmless cotton clothes as metal. Every time an innocent cotton-wearing person walked through the gates, the alarm would sound, and security guards would rush over to search that person. This would be an inappropriate security response. However, luckily the gates do have the technology to know when someone is not carrying any metal items. This is one role that Tregs fulfil – they help the system understand whether an incoming particle or cell is noxious or innocuous, thus allowing an appropriate immune response.

Essentially, the immune system has an army with a wide range of different attacking mechanisms and a type of military police (Tregs and others) to stop the army from doing things it should not do – just like a human army, which also has a military police force. An unregulated army is a dangerous thing! Microbial old friends and many friendly gut bacteria promote Tregs and so are intimately involved in maintaining the immune system's military police.

Graham talks about studies that have demonstrated the link between microbes and immunity. He mentions a study where researchers collected the microbes from a group of humans who were allergic to cow's' milk and then transplanted them into germ-free mice.[13] Germ-free mice are raised in a lab without any microbes, allowing them to be used in controlled scientific experiments. In this transplant study, the mice given the microbes from the allergic humans subsequently became allergic to milk. For comparison, the researchers inoculated another group of mice who were allergic to milk before the intervention with microbes from non-allergic humans. In this case, microbes from the non-allergic humans caused an appropriate immune response to the milk. In other words, the friendly microbes eliminated the allergy.

If you have allergies, imagine taking a pill that contains the microbes from a person without allergies. And now imagine it would cure your symptoms. This might not be as sci-fi as it sounds; researchers are working on similar concepts as I write these words.

In the late nineteenth century, microbiologist Louis Pasteur hypothesised that certain microbes were essential for the survival of so-called complex life-forms such as humans, other animals and plants, because these life-forms had co-evolved alongside those microbes. It seems he was referring to our microbial old friends without realising it. A century and a half later, his thoughts are truly substantiated.

The microbes in the guts of invertebrates, such as insects and worms, tend to be kept separate from the physiology of the host animal. They are kept either in specialised cells or in a proteinous and sugary mesh called chitin, pronounced 'kai-tin'. The diversity of microbes that live in these creatures is also relatively low. However, as mentioned earlier, when vertebrates first evolved on Earth around 500 million years ago, the microbial communities in animal hosts became far more numerous and complex. The microbes then started living very close to the host animal's physiology. Many microbes live in mucin (pronounced 'mew-sin') within the body. They are fed by this proteinous substance. From one perspective, our immune system is farming microbes from the environment. Graham says he likes pointing out that as humans, we have parallel traits with plants: 'We have a form of soil in our guts. We even call it "soil" sometimes ... and the microbes in that soil are feeding on the carbohydrates that our guts secrete and offer goodies that benefit our physiology.'

He goes on to say, 'This is like a plant, but the main difference is that the plant's soil is in the ground and not in an internal tube, like our gut.'

As humans, we are generalists. We will eat almost any non-toxic food we can get our hands on. Consuming these diverse foods, which come from diverse environments, means our guts are bombarded with an abundance of different microbes. The mammalian immune system had to develop a cool trick to deal with all this complexity – that is, to differentiate between the goodies, the baddies and all those in between. This particular trick was the evolution of the *adaptive* immune system, the complementary partner to the innate immune system.

The adaptive immune system

First, think about a cookery book. You've created a delicious recipe from scratch. But it's quite difficult to get right, involving a host of ingredients and a complex series of steps that must be undertaken in the right order, not to mention at the right temperature. If you don't write it down, you'll have to re-learn the recipe next time – a slow and cumbersome process that's prone to mistakes. However, if you record it in a recipe book, you can simply locate the page and replicate the recipe smoothly and with few errors. This is similar to the adaptive immune system.

The adaptive immune system has a recipe for dealing with pathogens, but it also remembers what they 'look' like so it can mount an efficient response when it encounters them again. Mutations are one of the adaptive immune system's closest allies. Indeed, the adaptive immune system produces a vast repertoire of 'receptors' by mutation. These receptors allow the adaptive immune system to recognise pathogens with similar patterns and deal with them effectively. In addition to this system of mutations, the immune system needs to select a repertoire of immune cells that match the microbes that appear in the local environment. That's why it's essential to be exposed to a diverse range of microbes from the first stages of life. This allows our bodies to produce tiny armies of 'memory cells' that maintain a record of all the pathogens our bodies encounter – much like recording our recipes in the cookbook. It also enables a rapid and effective immune response to similar pathogens in the future – just as our cookbook enables us to rapidly reproduce a meal, we created in the past.

Unlike the innate immune system, the adaptive immune system is highly specific. Once the immune system develops this repertoire of memory cells, it can recognise a bona fide invader. *Memory T cells* remember the pathogen. They act rather like USB flash drives, storing important data. *Memory B cells*, meanwhile, remember how to make the immune cells that render the pathogen useless.

During a pathogenic invasion, immune cells engulf and destroy the pathogen, and present it to *T helper cells*. You can think of these as the nightclub bouncers of the immune system. The T helper cells then seek out the other invaders and coordinate a full-body attack. They send a signal to *macrophages* and *killer T cells* – the ultimate warriors of the immune system. From here, you can probably guess the fate of the invader. You don't need to remember all these names, but just know that the immune system has an army with many different military personnel. They all play vital roles in detecting and eliminating an invading pathogen.

Both the innate and adaptive immune systems work closely together to coordinate a response to pathogens. There is a degree of cross-communication between them and our microbes, immune system and brain. These interactions all determine how our bodies and minds respond to infections. I'll return to this in Chapter 5.

During my conversation with Graham, we also chatted about *biofilms*. These are groups of cooperative microbes. Their cells stick to each other and to surfaces. The microbes are eventually embedded within a slimy sugary matrix.

In the case of bacteria, they do this firstly by communicating with each other. Bacteria use special molecules to detect the density

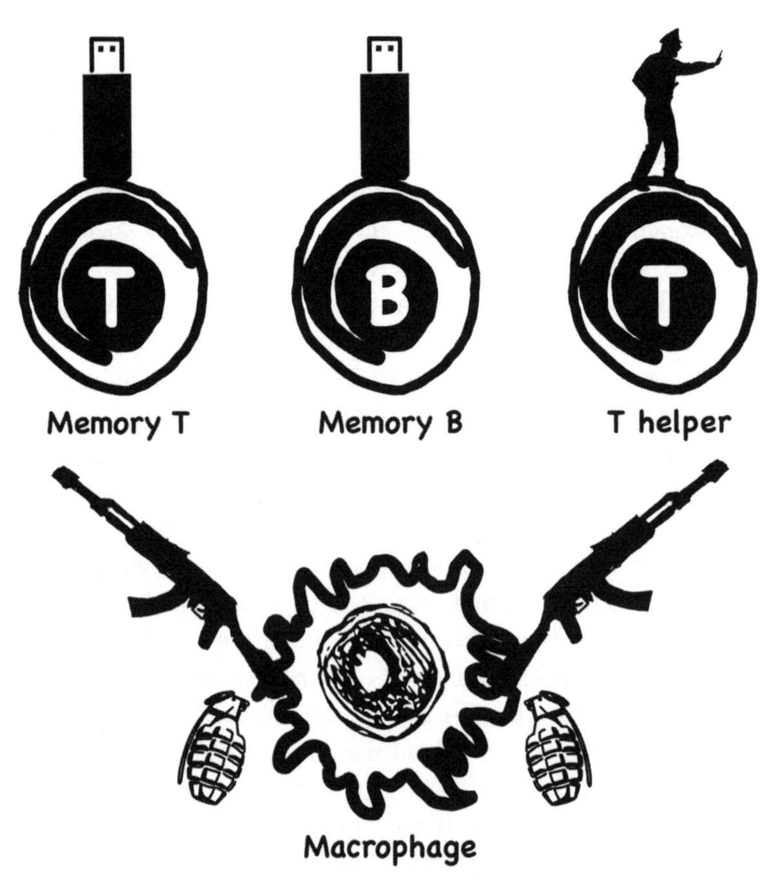

Immune cells (the USB drives, bouncers and ultimate warriors).

of other bacteria around them (because of course, they don't have eyes). When the cells reach a certain threshold, they can start to build these biofilms.

But why do they form biofilms? They do this to mount a group-based response – for example, to secrete a potent toxin, gather food or establish colonies that produce light (also known as bioluminescence). Essentially, a group-based action is far more potent than an individual action, so they stick together, literally. Biofilm formation makes it harder for our immune system to attack pathogens – it can be like a force field surrounding a group of bullies. But evidence shows that if you have a very complex community of microbes in your gut, it's less likely that any of them will succeed in forming a pathogenic biofilm. This is another reason we should nourish our microbiomes.

Graham and I swiftly returned to our conversation about the importance of microbial diversity in the human body. We discussed a final reason why it's essential to have a diverse community of microbes in our bodies: because there is a greater chance that a friendly microbe will occupy an ecological niche that a pathogen might otherwise try to steal. We refer to this broader concept of outcompeting others to occupy a given niche as the *competitive exclusion principle.*

We talked about microbes and the Western lifestyle. Some chronic disorders are due to persistent background inflammation in the body in Western high-income countries. These conditions are striking when considered in terms of high-income versus low-income countries and urban versus rural populations, particularly at the lower end of the socioeconomic scale. It's pretty easy to collect human blood samples to assess chronic inflammation.

If we take blood samples from people in many low-income countries, we will likely find low levels of cells that indicate chronic inflammation. The cells rise when a person has an infection, and then they drop to normal levels when the infection is over.

However, in many high-income countries, for example the United States, there is a persistently high level of chronic inflammation. This inflammation is linked to various conditions, including neurological issues such as depression. Scientists think the differences in inflammation between high- and low-income countries are attributed, in part, to gut microbes.

As we've discussed, microbes play essential roles in controlling inflammation. Many people consume ultra-processed foods in Western countries, with very little natural fibre and fewer naturally diverse microbes.[14] We're also surrounded by a concrete jungle, living and working in our giant synthetic cubes. Many people are less emotionally connected to nature; thus, they don't feel like regularly venturing out to woodlands, meadows and rivers. We're also wolfing down high levels of antibiotics that destroy our internal microbial ecosystems, like napalm to a forest.

I've spoken to Graham before about how differences in exposure to microbial diversity are linked to community health outcomes. A fascinating example in the United States involves two groups of people, the Amish and the Hutterites. These populations are incredibly similar genetically, geographically and culturally. However, a key difference between the two groups is their farming practices. The Amish still practise traditional methods, using hands-on approaches with more contact with animals, whereas the Hutterites have moved towards modern, mechanised and chemical-based farming. Researchers noticed that in addition to the differences in farming practices, they also differed in aspects of their health. For instance, the Amish population had a much lower prevalence and severity of inflammatory diseases such as asthma. Could this have something to do with the different farming approaches? Researchers dug a little deeper.[15]

They found significant variation in the faecal microbiomes (i.e. microbes in their poo) between the two populations. They also sampled the dust in the Amish and Hutterite houses to explore the environmental microbiome of their homes. Once again, they found significant differences in the microbial communities. They hypothesised that these differences might be due to the farming practices.

Mechanical and chemical methods likely reduce contact with diverse microbial communities, particularly those dwelling in the soil, plants and livestock. The researchers went a step further. They took the microbes from the house dust of each population and transplanted them into germ-free mice. They found that only the microbes from Amish house dust were protective against asthma in the mice.

But perhaps this was just the microbes from the house dust. How do we know they colonise the guts? Well, researchers went another step further. They collected stool samples from both human populations and found that gut microbes from Amish but not Hutterites protected the germ-free mice from asthma. There is quite a large body of evidence to suggest that growing up on a farm is protective against allergies. However, it seems that the type of farming practice makes a big difference. What is essential to our gut health is a farming system that promotes closer contact with our invisible friends.

Another fascinating study is that of the Finnish and Russian Karelians, carried out by Tari Haahtela and co-workers and published in 2015.[16] In ancestry and geography, these are two very similar populations. An area that supported both populations was divided in 1944, during the Second World War. The Finnish Karelians moved to the Finnish side and the Russian Karelians to the Russian side. As Haahtela said, 'the border between the Finnish and Russian Karelia marks one of the sharpest boundaries in the standard of living and health in the world. The adjacent areas are socially and economically distinct but similar in geology and climate.'

After the war, Finland experienced rapid economic and urban growth. This created the striking socioeconomic gap between the two populations we see today. During the Soviet era, the Russian Karelians were almost completely cut off from the outside world. On the Russian Karelian side of the border, people's living conditions resemble those that existed in Finland in the mid-twentieth century. Indeed, unlike the Finnish population, the Russian Karelians still lived in the countryside, kept farm animals and produced much of their own food. The natural environment was relatively well preserved.

Researchers noticed that levels of inflammatory diseases were different between the populations. The Finnish Karelians had far more pollen and pet allergies, particularly those born after 1944

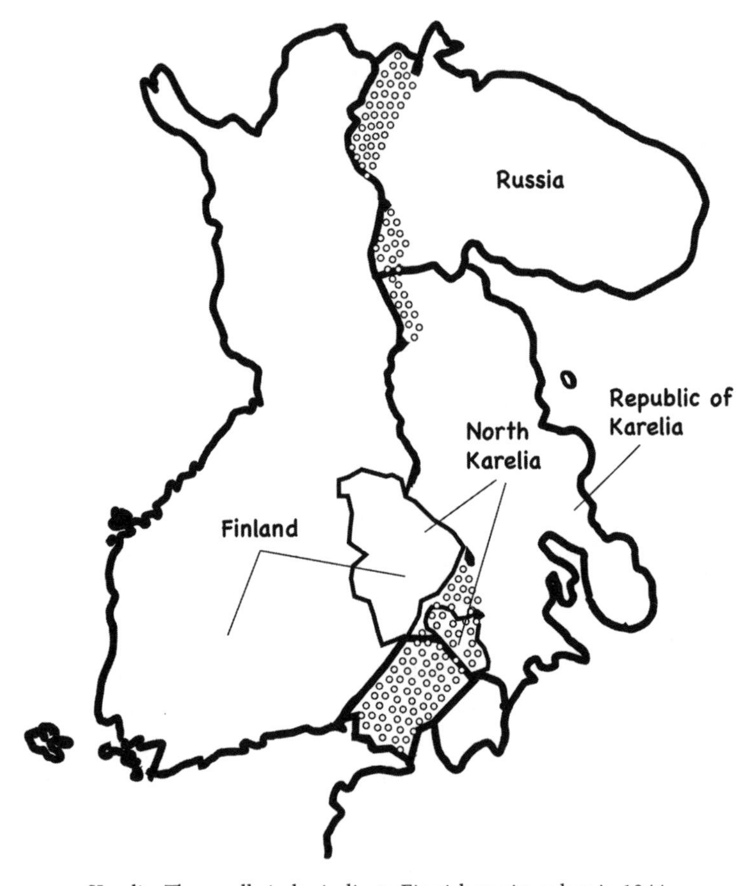

Karelia. The small circles indicate Finnish territory lost in 1944.

when the populations were divided. The researchers discovered that microbial communities in house dust and drinking water conferred greater allergy protection in the Russian Karelians. This supports the notion that our health will likely suffer if we alter the natural environment and our exposure to diverse microbes. This was supported by other studies in Finland showing that healthy children were more likely to live in biodiverse environments (those with more plants and animals) and had greater microbial diversity on and in their bodies.

Speaking of which, a fellow microbial ecologist, Dr Marja Roslund, and her team from Finland recently published a fantastic study.[17] Marja showed that by bringing local forest materials into children's schoolyards, the team were able to change the skin and gut microbiome of the children. Not only this, but the changes were associated with enhancements in the children's immune systems. They also assessed children's microbiomes and immune markers in standard urban schoolyards as a control measure.

Marja and her team also did a follow-up study two years later to test for long-term effects. Incredibly, they found that the forest materials in the schoolyards caused long-term shifts in the beneficial microbes among children. Keep an eye out for subsequent Finnish microbial ecology studies – I'm sure they'll have exciting results.

As most conversations naturally do in life, Graham's and mine drifts on to other matters; before long we find ourselves discussing the topic of eating soil. He tells me about anthropological studies performed in Africa a long time ago. The researchers recorded societies where it was fairly normal to eat soil, and children were eating dirt right up into their teens and in relatively large quantities. Recent studies have revealed that pregnant women commonly eat soil sticks sold in the markets of Central and East Africa.[18] These are known locally as 'pemba' and 'kichugu'. The sticks are made of soil from the walls of houses, termite mounds and the ground. The act of eating soil is known as *geophagy* – from the Greek *geo* ('earth') and *phagein* ('to eat'). Geophagy has a surprisingly long history around the world. In some countries, you can buy packets of soil in a range of flavours, including black pepper and cardamom, and in some grocery stores it is known as 'pregnancy clay'. In theory, it is possible that consuming soil will bring some benefits, including beneficial microbes and micronutrients. However, many soils now contain high levels of toxins and heavy metals because of human pollution. So any evolutionary advantage to geophagy may be overridden, particularly in areas where human disturbance and industrial activities are high.

Graham points me to a study in which researchers spent many hours watching the hygiene behaviours of human infants without intervening. It turns out the infants were getting through 20 grams of soil a day. Some kids gobbled up eleven handfuls of soil and ingested chicken faeces twice in six hours. It was estimated that a one-year-old infant ingesting 1 gram of chicken faeces and 20 grams of soil in a day would consume between 4 million and 23 million *E. coli* bacteria.[19] Graham says eating the chicken faeces was probably not a good idea. But it may be a different story for the soil – unless it's polluted.

Essentially all vertebrates seem to eat soil, especially in very early life. Graham jokes, 'Freud had the idea that it was something to do with loving your mother … you know, with Freud being a bit of a lunatic, but he was such a good writer that people believed what he said. Nevertheless, it's likely, at least partially, to do with spores.' Graham is talking about spore-forming bacteria. Much like plants develop seeds for dispersal, some bacteria grow spores – just like the *Streptomyces* I mentioned in the Introduction. The number of microbial species in the gut known to be spore-forming is not enormous. But, by some estimates, 60% of the entire bulk of individual organisms in the gut are thought to be capable of forming spores. The thing about many spore-forming microbes is that they're *anaerobic*. This means oxygen is toxic to them. The mother has difficulties transferring them to the baby because they just don't survive wafting around in the home. Graham says the answer is probably to pick them up as spores from the natural environment. This also means you can get a top-up if quantities are low, just by visiting the natural environment, the dirt, the biodiversity beneath our feet.

'There's the interesting point, that once upon a time we were all surrounded by human poo as well as the poo of other animals. Whereas nowadays, that's the one sort of poo that we don't meet because of the remarkable efficacy of the water closet and the fact that most people tend to wash their hands afterwards. So, human poo is floating away to who knows where in the sewers and is dealt with. At the same time, we're surrounded by the poo of every other conceivable furry or feathered, jumping, flying creature in the natural world. So, I have been wondering whether we're picking up gut microbes predominantly now from non-human species and whether this is important.'

This seems disgusting, but it's a great point. How much have the flushing toilet, the drains and the sewage systems changed our gut microbiomes over time?

Graham mentioned an old study on green iguanas, where the babies hatched and the first thing they ate in the nest chamber was soil.[20] The iguanas need to establish populations of microbes in the hindgut before digestive activity can begin. It's a rudimentary microbial fermentation system that supports faster growth and presumably helps degrade the vegetation more efficiently. After the first few weeks of life, the iguana hatchlings travel into the forest and obtain a more complex and effective gut microbial community by consuming the faeces of their seniors. Yum!

However, the classic example must be Australia's koala.[21] Very few other animals eat highly toxic eucalyptus leaves, so the koalas have an ecological niche all to themselves, just as long as they can detoxify the leaves and digest them. For this neat little evolutionary strategy to be successful, the koalas require help from microbes that can process the toxins. The fantastic thing is that they keep these microbes in the family by, yes, you guessed it – consuming parental poo. A joey (the young koala) consumes a liquefied form of the mother's faeces, known as 'pap'. This supplies the necessary microbes to detoxify and digest the eucalyptus leaves.

Our video call draws to an end, and Graham semi-jokingly wonders 'whether, if we ate koala poo, we'd be able to eat eucalyptus leaves – well that's up for debate'.

Some microbes are pathogenic, but far more are vital to our immune systems, keeping us healthy and ultimately alive. It took our fleet-footed, soil-dwelling mammal ancestors millions of years to develop our co-evolutionary relationship with these helpful microbes. Nowadays, many people's immune systems seem to be weakening, and we turn to antibiotics for help. But as we'll see in the next chapter, this is not always a healthy avenue for our bodies or landscapes, especially when antibiotics are misused. It turns out we may need to turn to some invisible friends such as bacteriophages (the viruses that hunt down bacteria) to help us out of this conundrum.

CHAPTER 3

Antibiotic-Resistant Landscapes

'Where there is power, there is resistance.'
—Michel Foucault

The vivacious hubbub of the city is a recent fledgling of the universe, evolutionarily speaking. Cities of at least one million people only arose during the latest 1% of human history. If we take the starting point as when hominins first began to use tools, these cities emerged during only the latest 0.07% of our history. Urbanisation exploded in the blink of an eye, and most humans now reside in landscapes quite distinct from those of their ancestors. Indeed, even during our lifetimes, our landscapes have constantly evolved. Human activities shift and shape the world around us, and the changing climate adds to this rapid transformation.

However, in addition to changing our landscapes, we are also changing our microbial land-, water- and air- 'scapes' or *microbioscapes*. Myriad communities of microbes dwell in cities – locations that were once forests, lagoons and grasslands. Microbial communities change as the landscapes change. The genetic make-up of the microbes also rapidly changes, and often in response to human activities. This quick-fire evolution is quite normal for microbes. They have short *generation times*, some reproducing in minutes. Yet, on average, it takes humans over 20 years to grow and have children. Greenland sharks, meanwhile, don't reach sexual maturity for 150 years – that's twice the average life expectancy of humans.[1] Five successive generations of humans could live to the current average global life expectancy of 73 years before a Greenland shark even reaches sexual maturity. And just think how many microbial generations could roam the Earth in 150 years. This rapid reproductive capacity and ability to swap genes with one another mean that microbes can adapt, and fast.

The ability to swiftly adapt can be fantastic for microbes. When ecological conditions change, no problem – so can they. This adaptability can also benefit humans. For example, we depend upon microbes and their myriad vital roles in our ecosystems. If they were unable to rapidly adapt, our ecosystems – along with our health and our well-being – would most likely suffer. However, this quick-fire evolution also means that microbes can adapt to antibiotics, a process known as 'antimicrobial resistance'. Unlike humans who pass on their genes via sexual reproduction (from parent to offspring; known as *vertical gene transfer*), some microbes are able to swap their genes with other microbes without having sex (known as *horizontal gene transfer*). Some of these newly acquired genes allow microbes to resist antibiotics. Bacteria continuously swap around swarms of antimicrobial-resistance genes far across our landscapes. Unfortunately, this poses a considerable threat to humans because we need antibiotics to fend off nasty infections.

Now the following is an important point: we use antibiotics to kill unwanted microbes, but microbes are also the source of many antibiotics. That's because some microbes have been naturally producing antibiotics for millions of years to defend themselves against other microbes. It's worth noting that the beginning of antimicrobial resistance was recently traced back to around 350 to 500 million years ago.[2]

Bacteria in the phylum Actinobacteria produce antibiotics in the soil. Actinobacteria produce these compounds to help defend themselves and outcompete other microbes in their environment. They use them as weapons. And they were likely doing this even before dinosaurs roamed the planet. Scientists have discovered that Actinobacteria also evolved genes that confer resistance to the antibiotic compounds they themselves produce. In other words, they evolved to protect themselves from their own weapons![3] Yet human exploitation of antibiotics in medicine and agriculture has shifted antibiotic resistance from the often-innocuous host producers to the disease-causing bacteria in just a few decades.

A recent study, however, showed that a specific type of the methicillin-resistant *Staphylococcus aureus*, also known as the MRSA superbug, actually evolved in the wild. The selective pressure of antibiotic production by a pathogenic fungus on the skin of European hedgehogs is thought to be responsible for this resistant strain of bacteria. These strains emerged in the 1800s, way before the modern-era use of clinical antibiotics. This discovery of wild resistance genes in human pathogens only strengthens the call to use any new antibiotics with care.[4]

But not all *Staphylococcus* are bad for our health. In fact, 20–40% of healthy humans have populations of *S. aureus* (the one that causes the superbug infections) living on our skin, so it's actually a common member of our skin microbiome. One reason *S. aureus* doesn't cause regular issues in healthy individuals is that it has some friendly cousins, *S. epidermis* and *S. hominis*; well, friendly to us, at least. Recent research has shown that *S. epidermis* and *S. hominis* produce a previously unknown antimicrobial compound that reduces *S. aureus*'s ability to replicate, thereby keeping *S. aureus*'s growth in check. Researchers found the numbers of these friendly cousins are depleted in people with atopic dermatitis, a skin disease linked to *S. aureus*. The researchers are now developing a therapeutic application. Dr Nakatsuji, who led the study, thinks that using antibiotics produced naturally by microbes in the skin is better than synthetic drugs because it doesn't kill the protective bacteria in the community: 'Antibiotic resistance is not likely to occur because the bacteria therapy is attacking pathogens by multiple different ways at once.'[5]

So we have several *Staphylococcus* species living in our skin microbiome; some may threaten our health, but others protect us from harm by producing antibiotics. Below is a slightly grainy image of the bacteria from my skin after I cultured them and viewed them under the microscope. They generally look like tiny poppy seeds. There are some black shapes and some lighter shapes. The black shapes are possibly *Staphylococcus*. It's difficult to see what species, but I hope they're a healthy mix of *S. epidermis* and *S. hominis*,

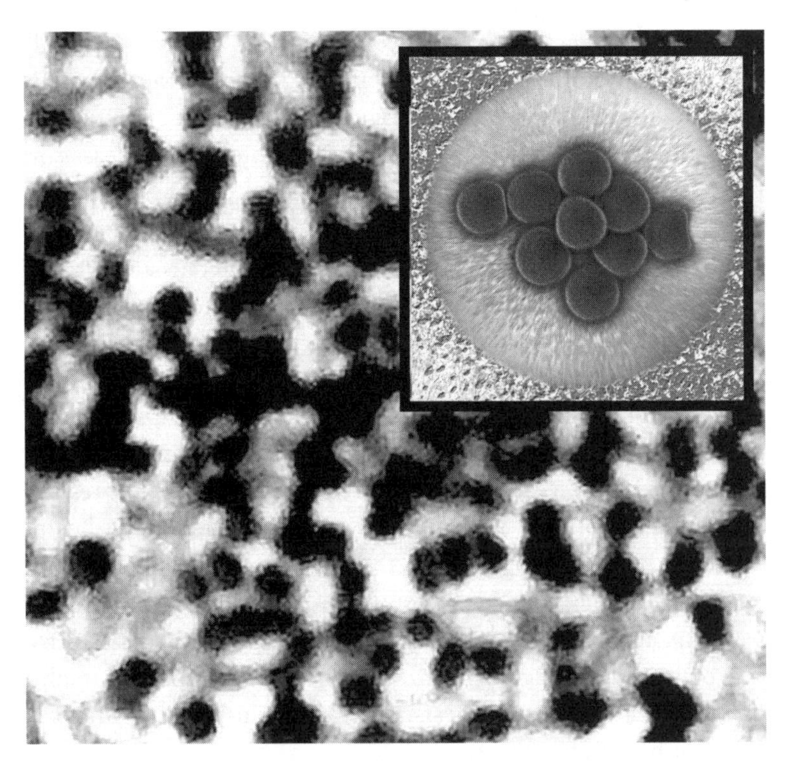

Bacteria from the author's skin overlaid with a clearer artistic impression of *Staphylococcus*.

perhaps using their antimicrobial compounds to keep *S. aureus* in check. The reason they're dark is that I stained the microscope slide. Staining is used to distinguish between two main types of bacteria – gram-positive and gram-negative – which have different cell wall structures. Gram-negative bacteria will turn darker because they don't have a thick outer layer of a sugary substance called *peptidoglycan*; hence, they absorb the dye. In contrast, gram-positive bacteria will turn a lighter colour because they do have this thick peptidoglycan layer and don't absorb much of the dye.

The 'modern antibiotic era'

Despite the popular narrative that the deliberate use of antibiotics by humans is confined to the post-scientific revolution era, evidence suggests we've been purposely using them for many hundreds of years. Tetracyclines – a group of potent broad-spectrum antibiotics – have been found in human skeletal remains from Sudan and Egypt dating back to at least 350 CE.[6] This has been tied to the unusually low rates of infectious diseases in the area at the time. Some researchers think the ancient people had mastered the science of fermentation and were purposely producing the antibiotic for therapeutic use. There is also similar evidence from Jordan and China. However, the 'modern antibiotic era' is typically associated with the likes of Paul Ehrlich (the late German physician, not the still-living American biologist) and Alexander Fleming in the early twentieth century. Incidentally, when Fleming serendipitously 'discovered' penicillin in 1928, it took a further 12 years to purify it sufficiently for clinical testing and an additional 5 years for its mass production to begin. Penicillins are a group of antibacterial drugs, and they come from *Penicillium* fungi. Again, the compounds are simply the fungus's weapons against other microbes. In Fleming's 1945 Nobel Prize acceptance speech, he warned that the overuse of penicillin might, one day, lead to bacterial resistance with spine-chilling implications for human health:

> But I would like to sound one note of warning. Penicillin is, to all intents and purposes, non-poisonous, so there is no need to worry about giving an overdose and poisoning the patient. There may be a danger, though, in underdosage. It is not difficult to make microbes resistant to penicillin in the laboratory by exposing them to concentrations not sufficient to kill them, and the same thing has occasionally happened in the body. The time may come when penicillin can be bought by anyone in the shops. Then there is the danger that the ignorant man may easily underdose himself, and by exposing his microbes to non-lethal quantities of the drug, make them resistant. Here is a hypothetical illustration. Mr. X has a sore throat. He buys

> some penicillin and gives himself not enough to kill the streptococci but enough to educate them to resist penicillin. He then infects his wife. Mrs. X gets pneumonia and is treated with penicillin. As the streptococci are now resistant to penicillin, the treatment fails. Mrs. X dies. Who is primarily responsible for Mrs. X's death? Why Mr. X, whose negligent use of penicillin changed the nature of the microbe. Moral: If you use penicillin, use enough.[7]

Yet there's another equally dangerous driver of antibiotic resistance. Agricultural and medical waste contain high concentrations of antimicrobial compounds – toxic concoctions, menacing powerhouses of evolution. Humans started using antibiotics in food production shortly after their modern-day mass synthesis in the 1930s. Antibiotics were used to treat and prevent diseases and to preserve food. The majority of people initially saw the rapid diffusion of antibiotics into each corner of the food industry as a life-saving revolution. By the 1950s, however, an increasing number of scientists became vocal about the risks of antimicrobial resistance. At the time, consumers were shocked to learn that 10% of milk in the United States was contaminated with penicillin.[8]

Meat consumption began to rise in the mid-twentieth century. In 1961, the average global annual meat consumption per capita was 24 kg. Today that figure has nearly doubled to 43 kg. This hides some hair-raising global inequities. For example, this figure is 100 kg in Australia and 120 kg in the United States.[9] Whereas in Bangladesh, the average annual consumption is just 4 kg. The rising global consumption of meat begets a wave of antibiotic use. Inoculating animal feed with antibiotics can raise productivity by reducing the expense of caring for sick animals. The rise of antibiotics in agriculture also spurred the dawn of industrial 'factory farms' where thousands of animals were (and still are) crammed into lots to increase productivity. Between 1970 and 1978, antibiotic use in the United States surged from 3,310 to 5,580 tonnes.[10] Antibiotics were also used in fish farms; 80% of these antibiotics would enter marine environments, resulting in the selection of antimicrobial resistance in wild fish populations. Meanwhile, an increase in plant diseases led to the expansion of antibiotics into the horticultural and food-crop industry, where European orchards would be sprayed with antimicrobials en masse. In the 1980s, concerns about chemicals on food products spawned the advent of the organic food industry. However, despite this mini-revolution, antibiotic overuse continues today. Indeed, a 67% increase in global agricultural antibiotic use is projected between 2010 and 2030, particularly in low- to middle-income countries where meat consumption is set to rise.[11]

The antibiotics given to farm animals spread across our landscapes primarily through defecation and urination. This process can spawn superbugs that become resistant to antibiotics. When humans consume meat, they can also consume the superbugs, and as Fleming warned us, a key concern is that diseases will arise that are effectively untreatable. Unfortunately, those who follow a plant-based diet might not be safe from this threat either – depending on where and how your food is grown. Indeed, the animal manure spread across fields to fertilise crops may also contain antimicrobial-resistance genes and superbugs. This means the vegetables could be covered in these not-so-friendly bacteria. Many common bacteria in non-human animals, like *Salmonella*, can also infect humans. Drug-resistant versions pass to humans through the food chain and then spread via international travel and trade. Because of environmental transmission, today's patients often already have antibiotic-resistant bacteria in and on their bodies without prior drug treatment.

Another important source of antibiotic-resistance genes in our landscapes is sewage. Water from human activities collects at waste-water treatment plants. Sadly, these are unintentional hotspots for antibiotic-resistance genes. In an ideal world, the processes at the waste-water treatment plants would neutralise antibiotic-resistant bacteria. However, they are not designed to do so. Therefore, once the treatment process is complete, resistant bacteria still make their way into the local environment through rivers and reservoirs.

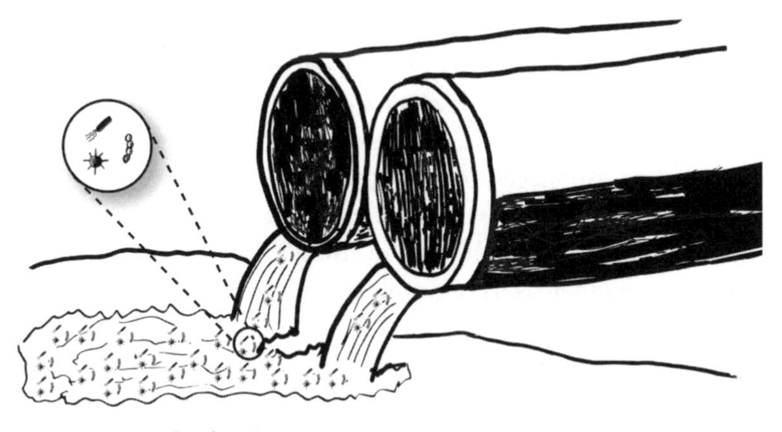

Armada of antibiotic-resistant bacteria in sewage discharge.

This conjures up rather poignant memories of being warned against swimming in the seas and rivers around the UK. Astonishingly, water companies in the UK can still let raw sewage spill into the rivers and seas after extreme weather events, which is often just slightly heavy rainfall. The Victorian sewage system in the UK is mostly outdated and fails to cope. Astoundingly, there were 403,171 spills of sewage into England's rivers and seas in 2020.[12] This equated to more than 3.1 million hours of, often raw, sewage spilling into our natural habitats. A recent report from The Rivers Trust revealed that more than half of the rivers in England are in 'poor condition', and none are in 'good overall condition'. We can attribute this primarily to the raw sewage spills, agriculture chemicals, and waste running off fields.

Just imagine the indomitable armada of antibiotic-resistant bacteria sailing in their fleets in unfathomable numbers through the pipes and into the rivers and seas when raw human sewage is discharged. This is the reality of the situation in the UK. Now think about the countries that lack sewage and waste-water facilities. The impacts on the environment and human health must be immense.

The fight against antimicrobial resistance

Fortunately, scientists are working on (a) controlling the use of current antibiotics, (b) creating new ones, (c) combatting resistance to new drugs and (d) finding other ways to fight microbial infections. Enlisting the help of unlikely invisible friends is one promising approach – viruses that can attack infectious bacteria but don't harm humans. As discussed in Chapter 1, we call these viruses bacteriophages, or phages. Researchers are assessing the effectiveness of these phages and determining the possible downstream ecological impacts of using them. After all, the last thing we want to do is fix one problem only to inadvertently create another.

Scientists actually discovered and developed phage therapy even before antibiotics were mass produced. Ernest Hankin, a British bacteriologist, reported the antibiotic activity of phages back in 1896.[13] They were eventually isolated in 1916 by Felix d'Hérelle. The cosmetic company L'Oréal was marketing phages as effective antibiotics even before the mass production of penicillin in the 1940s.[14] However, many of the earlier studies involving phages were published in non-English journals, so phage therapies were not vigorously pursued in the West. Their immense potential as an alternative to conventional antibiotics warrants greater attention.

Soil biodiversity

A recent study showed that the loss of soil microbial diversity could exacerbate the spread of antimicrobial-resistance genes.[15] This might be due to competitors and predators reducing the density of resistance gene holders. This could have important implications for human health by making infections harder to treat whilst increasing the risk of spreading diseases. It is possible that restoring or altering the microbial diversity in a given ecosystem could help curb the spread of these menacing genes – as suggested by another study into soil bacteria.[16]

If so, there is an imperative need to support and guard these native invisible friends. Some inordinately fascinating research from China recently revealed that one particular small, albeit visible, friend – the earthworm – may reduce the number of antimicrobial-resistance genes in soils.[17] The researchers collected earthworms and soil from 28 provinces in China. They analysed the organic contents of the earthworms and found that the number of antimicrobial-resistance genes was far lower in their faeces than in their guts. This suggested that the earthworms could mitigate and neutralise these genes during digestion. The researchers also found that earthworms reduced antimicrobial-resistance genes in the surrounding soil samples. Not only do earthworms improve the soil by aerating it, breaking down organic matter and converting nutrients into available forms for plants, they could be vital in the fight against the looming threat of antibiotic resistance. Soil invertebrates also harbour diverse and unique invisible friends. As the authors of the Chinese study said, the soil is a hidden source which is 'rarely considered in biodiversity and conservation debates, and stresses the importance of preserving soil invertebrates' such as earthworms.

We don't yet have a good understanding of the impact antibiotic-resistance genes might have on the human benefits of exposure to environmental microbes such as the *old friends* mentioned in the preceding chapter. But these genes are likely adding to the depletion of microbial biodiversity already taking place across our new landscapes. This, in turn, could be detrimental to our health and our ecosystems. As noted, scientists have linked the loss of soil microbial diversity, for instance through monocultures and habitat clearance, to the exacerbation of the spread of antibiotic resistance. Whether restoring our habitats and the diversity of invisible friends reduces this spread is yet to be elucidated. But this will likely be an essential research avenue in the future.

CHAPTER 4

Microbes and Social Equity

'Not everything that is faced can be changed, but nothing can be changed until it is faced.'
—James Baldwin

Can we view microbes as a facet of social equity? How do microbes, ants, pollution and worms relate to communities having access to different opportunities and outcomes? I'll come back to this later in the chapter.

Sometimes it's good to take a step back and view our social world through an ecological lens. By doing so, we often find seemingly unlikely connections between social and ecological entities that might otherwise go unnoticed, such as how microbes relate to equity in health and happiness. I have spent much of my time in academia combining microbial ecology research with social research (the study of trends, behaviours and dynamics commonly displayed by people and societies). This chapter draws upon my own experiences and thoughts on these interconnected fields, along with the views of my colleague from the United States, Dr Sue Ishaq, who is a leader in the emerging field of *microbes and social equity*.

But first, what is equity, and what's the difference between equity and equality? There's an important distinction between these terms. We all have different needs, and *equality* assumes everyone will benefit by having access to the same resources. *Equity*, on the other hand, recognises that each individual or community has different circumstances requiring different opportunities and resources to achieve equal outcomes. To illustrate this, I've redrawn a famous sketch below. It shows a tall person, a short person and

Equality versus Equity.

a person in a wheelchair, all trying to see over the fence. On the left (representing equality), they all have equal access to the same resource – a sturdy box with identical dimensions. But only the tall person can use the box to see over the fence. The small person is getting an eyeful of fence panels, and the person in a wheelchair can't even use the box. Now take a look at the right-hand image (representing equity). We give each person appropriate resources (a small box, a tall box and a wheelchair ramp) to achieve equal outcomes. They're now all able to see over the fence. An example of equality in a 'microbes and social equity' sense would be all grocery shops stocking 'healthy' foods that nourish the gut microbiome. Whereas equity would ensure that everyone from all socioeconomic backgrounds could afford these healthy foods.

Our social world is peppered with disparities in opportunities and outcomes, and they're often related to the condition of the environments we live in. We should all be untrammelled by these inequities. Yet many communities endure immoderate and dangerous exposure to air, noise and light pollution, along with localised heat stress or the 'urban heat island effect'. This is particularly the case in poorer communities – those with a low socioeconomic status. These stressors take a toll on the health of humans and the other organisms in the surrounding ecosystems. The broken system is often characterised by the poor inclusion and distribution of natural environments (e.g. woodlands, meadows, lakes) and biodiversity in deprived urban areas. When these environments are present in deprived areas, they're often blighted by industrial pollution, engulfed by chemical clouds that spew from factory chimneys; they're also often unsafe and inaccessible. Moreover, some residents in deprived areas feel trapped and downtrodden, and often don't fulfil their potential in the education, health and employment realms. One study suggested children living in the poorest communities in the UK were 4.5 times more likely to suffer from severe mental health issues than children in the most affluent communities.[1] Turning to drugs and alcohol is a common way to escape reality. But this only exacerbates the health issues, which worsen the employment issues. Round and round it goes in a vicious cycle.

Disparities in living conditions and the quality of environments are often linked to structural classism (discrimination on the basis of social class) and racism (discrimination on the basis of race). This is exemplified by the poorer quality of life that Indigenous Peoples and other minoritised groups often endure. It's also highlighted by the apparent disparities in the quality of environments dictated by racial segregation in many cities. These disparities underscore the concept of *ecological justice* – defined as the fair distribution of environmental benefits and burdens. Social equity, meanwhile, is impartiality and fairness for all people. It considers historical and current inequities to ensure the entire community has access to similar opportunities and outcomes.

Do disparities in living conditions, natural environments and community well-being affect exposure to microbes – both the friends and the foes? Do they affect microbial ecosystems that, in turn, affect society?

I am tremendously lucky to be familiar with the person 'in the know' – an inspiring lecturer and researcher from the University of Maine in the United States, Dr Sue Ishaq, mentioned earlier.[2] I've had the pleasure of working with Sue on several publications, and she kindly shared her thoughts on microbes and social equity for this chapter.

Sue first began using the term 'microbes and social equity' around the winter of 2018. It sprang from conversations she had with colleagues and developed out of necessity. At the time, Sue chatted to architects about their concepts of ecological justice, and her concepts of the microbial communities found indoors and their impacts on occupant health. As usual in science, there were far more questions than answers about the microbes found indoors. Indeed, we still haven't comprehensively answered 'What constitutes a healthy microbiome in a building?'. In considering questions like this, she drew upon her gut microbiome training and knowledge of how microbes help us digest our food and generate valuable chemicals that sustain animal health.

Sue used a food-related analogy: 'It doesn't matter how many health-promoting benefits fresh fruits and vegetables provide, if there simply aren't any available to you or any within your budget.' This concept of limited access to nutritious food is also known as a 'food desert'. One potential solution would be to open more grocery stores with government subsidies on food near the people who had none before. This speaks to equitable access to resources through the microbes and social equity lens. Indeed, having access to various dietary fibres allows a person to recruit diverse microbes to their gut, which confer health benefits. This is much like how having access to biodiverse environments (natural areas with lots of plants, animals and microbes) enables a person to recruit many different microbes into their bodies – their walking ecosystems – particularly in early life.

Sue began teaching her students how lifestyle and environmental conditions could alter the microbiome and impact people's health. This process culminated in a 10,000-word publication featuring all the students.[3] When preparing the publication, Sue's team struggled to find relevant literature around these concepts because microbiome papers rarely mentioned 'equity' or 'justice'. Therefore, Sue wasn't expecting anyone to reach out to her about next steps. Nonetheless, after publishing the paper several people made contact, myself included. One thing led to another and the Microbes and Social Equity working group was established, which Sue continues to lead.[4]

I asked Sue, 'What are some of the broad implications of working on microbes and social equity?' She explained that so many aspects of our lives are connected to microbes somehow, even just in passing; thus there are many implications of studying the relationships a community of microbes might have with people and vice versa.

Access to a varied and healthy diet, especially fibrous foods, is vital to the integrity of our body's microbiome. Still, many people do not have access to these quality foods. People in deprived areas are far more likely to eat cheap, ultra-processed foods that are detrimental to their bodies and microbial denizens.[5] This underscores a critical social injustice. Preventative healthcare services are also vital, particularly dental and reproductive healthcare. Indeed, oral and vaginal microbiomes can be overlooked despite their connection to whole-body health. Quality natural environments are essential too, but again, many people simply do not have access to these life-supporting resources.

A community's habitat (living environment) is central to its health. Indeed, we are, in part, products of our environments. Several studies have revealed that the distribution of urban green spaces (e.g. parks, woodlands, meadows) can disproportionately favour particular communities – for example, those with a higher socioeconomic status and those from white ethnic backgrounds.[6] Other studies suggest that the number of green spaces might not differ enormously between areas with higher and lower deprivation. However, the quality of the environments and access to them often vary dramatically.[7] Therefore, some communities may be less exposed to the diverse microbes of natural environments due to access and quality issues. Consequently, the potential benefits of exposure to health-promoting microbiomes in the environment may be unequally distributed. And then we have pathogens. Some habitats may harbour more pathogens; therefore, some communities will be more exposed to the nasty microbes than others.

Let's run a thought experiment. Some people live in cleaner villages with more natural features than others. Large veteran trees might line the streets, along with woodland copses and luscious meadows instead of industrial estates. Air, light and noise pollution are all considerably lower. The people in these communities most likely breathe in more diverse and less pathogenic microbes from the air, and shop at higher-end food stores. The people here almost live in an entirely different world to people in impoverished villages. The former might have huge houses with equally massive gardens, clean, safe and accessible green spaces, and access to higher-quality, microbiome-nourishing foods. Here we see a clear hierarchy in terms of the quality of living environments.

Another important factor to consider is a phenomenon known as *time poverty*. Many people in highly deprived areas work several jobs, are on zero-hour contracts or do a disproportionately high level of shift work. This inequity not only eats away at a person's health, but it eats away at their 'free time'. This lack of free time reduces a person's ability to engage in nature-based activities (such as spending time in biodiverse green spaces). It also contributes to a person's stress levels, which in turn exacerbate health issues.

These factors likely equate to a hierarchy of exposure to beneficial microbes, from the top to the bottom of the socioeconomic spectrum. They also equate to disparities in people's ability to nourish their microbiomes and, thus, their health. Therefore, we can indeed view microbes as a facet of social equity.

A phenomenon exists in the UK called the Glasgow Effect.[8] This refers to the lower life expectancy of residents of Glasgow in Scotland compared to the rest of the UK. *The Economist* wrote in 2012: 'It is as if a malign vapour rises from the Clyde at night and settles in the lungs of sleeping Glaswegians.'[9]

The Glasgow Effect also plays out within the city itself; that is, we see huge disparities in life expectancy between different areas of the city. Notably, in 2008, one report estimated that male life expectancy at birth in the Calton area of Glasgow was a mere 54 years.[10] However, it is now more likely to be in the sixties – still a very young age. Recent data suggests that men in the Greater Govan area of Glasgow can now expect to live for an average of 65 years.[11] If you hop over the M8 motorway to the Pollokshields area, you'll find that men can expect to live to the ripe old age of 83. This is a gap of 18 years in two closely neighbouring regions of the same city.

Scientists have put various hypotheses forward to explain this disparity, including land contamination by toxins, higher derelict land levels, and poor housing quality and social support. All these phenomena could potentially drive inequities in exposure to microbes. As a friend and colleague of mine, Dr Alan Logan, said in 2015, 'in Western industrial nations, a disparity of microbes might

A drawing by Susan Prescott reflecting on socioeconomic disparities and the environmental and commercial drivers of dysbiotic drift.

be expected among the socioeconomically disadvantaged, those who face more profound environmental forces’.[12] Alan termed this disparity of microbes *dysbiotic drift*. He highlighted how the barrage of stressors associated with urbanisation and commercialisation (e.g. pollution, crowding, acoustic stress, heat stress, Westernised dietary patterns and sedentary behaviours) also associated with marked shifts in gut microbes. Alan said, ‘it becomes plain to see that at virtually every theoretical turn in which dysbiosis could arise, the socioeconomically disadvantaged may be at higher risk’.

Commercial tactics in deprived communities are fundamental drivers of the structural poverty that affects microbial health and integrity. Indeed, the entire commercial environment in disadvantaged communities is engineered to promote the continued consumption of unhealthy, ultra-processed foods by default. Unfortunately, these foods are hugely detrimental to the gut microbiome, reinforcing the vicious cycle. Alan and his wife, the world-renowned immunologist Professor Susan Prescott, have written widely on this topic, so I encourage you to delve into the literature they have published in both scientific papers and books.[13]

Antibiotics and socioeconomic status

Evidence suggests that early-life antibiotic use can change the gut microbiome and predispose children to obesity and immune disorders. A recent study revealed that children from poorer communities received significantly more antibiotics early in life.[14] Given what we know about the importance of developing a diverse microbiome in early life (refer back to Chapter 2), this could be hugely detrimental to the robustness of the less affluent children's immune systems later on. Exposure to more pathogens due to poorer living conditions is probably one reason for the greater usage of antibiotics in less affluent children. However, it's important to point out that it might not be all rosy at the higher end of the socioeconomic spectrum either. For instance, evidence suggests affluent children might receive more antibiotics later in life, though the reasons for this are yet unknown.

Strategies to counter the negative health outcomes

We know that diet and access to quality green spaces can positively influence the human microbiome. But how do we maximise these benefits? One way is to increase opportunities for people to grow, tend and harvest healthy foods that promote contact with diverse microbes in natural environments. An example of this is a community garden. We can view gardening as a nature-based intervention. More on this later – but in essence, people gather and grow vegetables, gain an ecological education, socialise, exercise and of course, get hands-on with the soil. From a well-being perspective, these activities tick many boxes. They are fantastic ways to foster reciprocal relationships with the land. Gardening exposes participants to a diverse environmental microbiome (which is likely to be more important in early life as a person's core microbiome develops) while providing vital prebiotics in fresh fruits and vegetables.

As alluded to earlier, it is also important to consider ecological justice and social equity in the context of *pathogenic* microbes. For example, do specific environments harbour higher proportions of non-beneficial microbes? A colleague of mine in Australia recently found that human-disturbed or degraded land may favour opportunistic bacteria, including pathogenic species.[15] Another study found that human-degraded land can release pathogenic fungal spores. We also know that the dense urban jungle can prevent the transfer of diverse microbes to the indoor environment,[16] and indoor environments can harbour higher proportions of human-associated pathogens.[17] So, creating socially inclusive green spaces with lots of plants and animals might also help to reduce our exposure to pathogens whilst promoting our exposure to invisible friends. But again, not everyone has this opportunity.

Recent strategies to address these social equity issues include the growing movement of 'social and green prescribing'. *Green prescriptions* are also known as 'nature-based health interventions'.[18] They are typically given by GPs and are designed for patients with a defined need, but they could also be used proactively to support a healthy lifestyle. They have the potential to supplement orthodox medical treatments, particularly those aimed at addressing chronic diseases such as cardiovascular and mental health issues. We also think that green prescriptions could bring significant ecological, economic and social benefits by helping to foster behaviours that are nature-friendly, such as recycling, conservation and ethical purchasing. A green prescription might involve something like the community gardening I mentioned earlier (in this scenario, it is more systematic, and is sometimes known as 'therapeutic horticulture'). Alternatively, it could involve volunteering for a local biodiversity conservation group, or it could come in a much simpler form: walking in a forest and immersing oneself in the sights, smells and sounds of nature.

People refer to the latter intervention as 'forest bathing' – so called because when among the trees, you are bathing in nature and beneficial plant-based compounds called *phytoncides*, and diverse microbial assemblages that sail through the air. These tiny organisms bombard humans daily. Studies have shown that up to a million microbial cells can be found in a single cubic metre of air, and a person can inhale 100 million bacteria each day.[19] We want to be bombarded by the naturally diverse communities, not the opportunistic pathogens. We know phytoncides (plant-based chemicals) can enhance the human immune system. But it is not only the immune system that may benefit from these natural compounds. In a recent study using mice, phytoncides even interacted with chemicals and cells in the brain to improve the rodents' ability to sleep.[20] Being stuck indoors surrounded by artificial light and disconnected from our microbial friends could be contributing to sleeping problems.

Forest bathing is known as *shinrin-yoku* (森林浴) in Japan, where it was first popularised, and it's proving to be a popular green prescription. The opportunity to 'bathe' in friendly microbes and plant chemicals should be available to all.

As I write this from my camping chair in the pine forest on the hill, I take a deep breath and a moment to notice the natural contours and bewitching sounds around me. I am indeed forest bathing whilst writing about forest bathing, which certainly helps in describing the experience. Again, I am lucky to have this wondrous forest on my doorstep. However, these opportunities are not equally distributed. Improving the quality of people's living environments and promoting a (mutualistic) symbiotic relationship with nature must be a priority if we are to address social inequity issues.

The ecological lens

Whilst talking to Sue for this chapter, I asked her if she has any messages for budding researchers in the emerging field of microbes and social equity. 'Perhaps the most valuable lesson from my work is that we cannot assume what microbial communities will do in a situation or ecosystem which hasn't already been evaluated, even if we know what animals or plants would do.' This is important because we end up rejecting many hypotheses involving microbes. The social realm adds another layer of complexity that requires careful consideration. We need to bridge the gap between science and society so that they inform each other in another kind of mutualistic symbiosis.

Sue also looks at social equity issues through an ecological lens, which has led to some fascinating hypotheses, and here is where we get back to the ants and worms mentioned at the very beginning of the chapter. Sue tells me a story about the European fire ant (*Myrmica rubra*) to demonstrate how microbes can affect social equity in not-so-obvious ways. The ants are native to Europe, but in recent decades, they've been found in North America. They're now considered 'invasive' (although they were taken across the Atlantic by the ultimate invasive species – *Homo sapiens*). Being an invasive species, the fire ants harm the native ecosystems. They damage local ant populations and other insects and their ability to disperse native seeds, thereby impacting the food web. They also have a painful sting, which can cause considerable injuries in people who are particularly sensitive to their toxins.

The next character in the story is a nematode worm, *Pristionchus entomophagus*, found in the soils of North America. The name *entomophagus* gives us some clues as to the worm's behaviour: *entomo* from the Greek for 'insect', and *phagus* from the Greek *phagein* meaning 'to eat'. They eat insects. This particular nematode worm invades ants through the oral or anal cavity, but it doesn't kill them. The juvenile worm waits for the ants to die of other causes before they mature and eat them. This nematode also ingests soil bacteria and has a keen sense of 'smell', meaning they can seek out the bacteria they particularly like or avoid the nasty ones. Sue and her colleagues wondered if the nematodes could transfer pathogenic bacteria into the European fire ants. They studied collapsed fire ant colonies and found that the transfer of bacteria from the nematodes to ants could contribute to ant mortality and death.[21] We need follow-up studies, but it does point to a potential solution to control the invasive ant.

This story shows that understanding the dynamics of hosts and their microbes in ecosystems, and then zooming out to see the bigger picture, might help us remove invasive species that impact ecosystems and communities. The people who deal with the consequences of the invasive species are rarely those who introduced it in the first place. This concept also plays out in other realms. For instance, a recent study investigated how much air pollution is created 'on your behalf' versus how much you are exposed to, and revealed some stark racial-ethnic inequities.[22] The study showed that air pollution exposure in the United States is disproportionately caused by the 'non-Hispanic white majority' but also disproportionately inhaled by black and Hispanic minorities. As pollution affects the environmental and human microbiome, this is another example of how microbes relate to social equity. It also highlights the mythological nature of the idea that we're individually responsible for or have full agency over our health.

One thing is clear; we can indeed view microbes as a facet of social equity. They also form part of the recipe to, as my friend Dr Gillian Orrow would say, 'create the conditions for health to flourish'. A deeper consideration of microbial exposure and social equity in research, planning and policy is therefore imperative.

CHAPTER 5

The Psychobiotic Revolution

'We are all of us walking communities of bacteria. The world shimmers. A pointillist landscape made of tiny living beings.'
—Lynn Margulis

Many moons ago – at least 2,000 years – practitioners of traditional Chinese medicine recognised that the stomach, and its partner, the spleen, were strongly affected by anxiety and overthinking.[1] Chinese doctors were acutely aware of the connections between the mind, the body and the environment. And over 2,000 years ago in Kos, Hippocrates purportedly said, 'All disease begins in the gut'.[2] These examples highlight the ancient recognition of the gut's important role in human health and well-being. More broadly, holistic perspectives of health are rooted in Indigenous cultures. For millennia, Indigenous Peoples have viewed nature, including humans, as a densely tangled web of interconnected subjects and not simply a collection of discrete objects. We are 'extensions of the earth'.[3] After speaking to First Nations scholars, I've learnt that Indigenous Peoples have been aware of the deep interconnections between all body systems for many thousands of years. Each one of us is an ecosystem within an ecosystem. Everything interconnects, including the gut, the brain and our bodies, along with our surrounding environments. Recent scientific breakthroughs have revealed that all these phenomena are in some ways connected through nested layers of invisible biodiversity – the microbes inside and around us.

In Western societies, eighteenth-century physicians such as Robert Whytt (1714–1766) – the first physician to King George III – were arguably late to the show.[4] Still, they also recognised the connections between the inner body organs and the nervous system. Whytt observed that the gut possessed a complex web of nerve endings that dispensed 'nervous energy' throughout the body. In the early nineteenth century, many books were published on gut health, some describing how the stomach was 'the sensorium of organic life' or the 'great abdominal brain'.[5] These terms echo the idea that the gut influences the brain and behaviour, and vice versa. In 1829, London-based surgeon John Abernethy (1764–1831) published the *Abernethian Code of Health and Longevity or Everyone's health in his own keeping by the proper regulation of the stomach and bowels in order to the attaining and securing those invaluable blessings*.[6] In this book, Abernethy traced all physical and mental disorders back to what he termed 'gastric derangement'. He recognised that anxiety could reduce appetite and that impaired digestion caused sleep disorders, fatigue and 'lowness of spirit'. At the time, Britain became more industrialised, and with this, the popularity of refined foods crept up like a strangling vine of ivy on an unsuspecting tree. Abernethy understood that this type of diet was problematic, and he was a fervent proponent of eating simple, unrefined foods to sustain a healthy body and mind. In the years since, terms such as 'gut instinct', 'gut feeling', 'I've got butterflies in my stomach' and 'it takes real guts to do that' have made their way into our everyday vernacular, reflecting the importance of the gut–brain connection.

At the time of the *Abernethian Code*, our knowledge of microbes was rudimentary, although we had known about them for over a century. Indeed, back in the seventeenth century, Antonie van Leeuwenhoek (1632–1723) and Robert Hooke (1635–1703) discovered micro-fungi, protozoa and bacteria using very simple microscopes. By the 1860s, our knowledge was advancing and Louis Pasteur published his germ theory – describing how certain microbes can invade the human body and cause diseases. At this time, we presumed that microbes could enter the human body through media such as the air. Indeed, a surgeon from this era strangely described non-pathogenic microbes floating in the air as 'eunuchs',[7] which comes from the Greek words for 'bed' and 'to guard over' – referring to the original term that describes a castrated man employed to guard over a woman's living area in China and the Middle East. Presumably, people used the term 'eunuch' to describe the inability of some microbes to function as pathogens. Nonetheless, it wasn't until the early twentieth century that scientists began to think that the gut–brain connection might have something to do with the denizens of our guts, the microbes.

Ukrainian immunologist Élie Metchnikoff (1845–1916) was awarded the Nobel Prize in 1908 with the German physician Paul Ehrlich for his work on innate immunity. However, a year earlier, Metchnikoff hypothesised that replacing the number of 'putrefactive' bacteria – that is, decay-causing bacteria, and certainly not eunuchs – in the gut with lactic acid bacteria could improve bowel health and prolong life.[8] This was revolutionary thinking for the time, and eventually, some 50 years later, it inspired the term 'probiotics'. Today, we define probiotics as live microbes, which confer a health benefit to the host when administered in adequate quantities. Research into probiotics has boomed since around the turn of the millennium, and there are now signs that Metchnikoff's original hypothesis – from over 100 years ago – was indeed correct.

In 2005, a young neuroscientist was heading back home to Ireland. He had travelled the world, working in the pharmacology industry in Australia, the United States and Switzerland. Now a professor of anatomy and neuroscience, John Cryan grew up on a cattle farm near Galway on the rugged coastline of western Ireland. John was a first-generation college student. But despite these humble roots, he is now a leading figure in the world of neuroscience, and his interests lie in understanding the connections between microbes, the gut and the brain – also termed the *microbiota–gut–brain axis*. According to John himself, he benefited from the Irish education system, which encouraged a wide variety of approaches to teaching and learning, along with philosophical and intellectual independence. John's PhD supervisor, Professor Brian Leonard inspired him immensely. Brian was a fellow neuroscientist who always encouraged creativity and challenged his students to think laterally. Professor Leonard was interested in the idea of a gut–brain axis way ahead of his time, and he was a proponent of embracing alternative views. This encouragement helped John to break free from the reductionist chains that often stifle creativity in science. He started to think more holistically, to see connections everywhere and between everything. He decided to take his experience in pharmacology (the study of the uses, effects and modes of action of drugs and biochemicals on living organisms) and apply it to help understand the impacts of stress on the brain. Along with John's 'systems thinking' approach, this decision took him deeper down a rabbit hole, a marvellously entangled place where a vast assemblage of microbes, neurons, biochemicals and discoveries awaited.

I video-called John to chat about his work. He informed me that upon his return to Ireland in 2005, he went to work at University College Cork, enticed by a new state-of-the-art microbiome research centre. He continued to work on stress and the brain. He noticed that animals stressed in early life had poorer health in adulthood. In addition to this, the stressed animals had different microbes residing in their guts. This was a lightbulb moment for John. It showed that stress leaves its signature in the host's microbes. Excited by this tantalising finding, John and his team in Cork set out to design scientific methods to improve our understanding of the relationship between stress and microbes in animals. The team studied the vagus nerve and used *Lactobacillus* bacteria, which have anti-anxiety effects.

The vagus nerve connects the central nervous system to many organs around the body – including the gastrointestinal tract. They found that brains in germ-free rodents were seriously underdeveloped. As mentioned earlier, germ-free rodents are mice or rats raised in a laboratory, and scientists deliberately prevent them from having microbes. This allows them to be used in controlled scientific experiments. Germ-free rodents typically have major neurological and social issues, and therefore display increased anxiety and avoidance of social situations. Because germ-free rodents don't have microbes, this suggests that microbes affect brain development and social behaviours.

Scientists have since found that germ-free animals have deficits in fear learning and pain processing in addition to social anxiety.[9] John tells me this could have important implications for understanding abnormal social behaviours and even autism in humans. It seems that the social impact of our microbial residents can be significant. In a social environment, microbes pass from one person to another. And as I mentioned earlier, we can view ourselves as walking ecosystems, constantly exchanging microbes and other compounds with the environment around us. We emit an inordinate number of microbes every day, and each one of us has a signature microbial cloud following us around wherever we go. Studies have demonstrated that your microbiome will be much more like the microbiome of other people living in your house than the microbiome of people who live elsewhere.[10] We also exchange microbes with our pets.[11] So, the people and the animals around us influence our microbial make-up. This could affect our behaviour and inclination to socialise.

I began wondering whether our personalities may synchronise because of these exchanges in microbes. And strangely, whether the concept of menstrual synchrony (whereby women who live together sometimes observe their menstrual cycles synchronising) could be related to these exchanges in microbes. After all, microbes are essential players in the menstrual cycle – for example, by activating oestrogen. Our microbial friends also produce and secrete hormones inside us, respond to our hormones and regulate the expression of our hormones. My mind was heading towards the entrance to John's entangled rabbit hole.

Fortunately, John changed the direction of the conversation, which snapped me out of this whirlwind of thoughts. He told me that it's not just our environments that have a considerable effect on our microbes and behaviour; *time* does too. Interactions between our microbiomes and our brains are different in early life compared to adolescence.[12] And these adolescent interactions are distinct from those in adulthood. I wonder whether these differing microbe–brain interactions at different times in our lives partially determine our changing moods, personas and values over time?

John talks about how the COVID-19 social restriction measures, or 'lockdowns', could also affect our microbiomes, for example by preventing the exchange of microbes between ourselves and other people and between ourselves and our environments. I have seen far fewer people over the last 18 months, sometimes only exchanging microbial clouds with my wife and our cat. Will my health and behaviour somehow be affected by this later in life?

In summary, we know microbes can 'talk' to the brain. And reciprocally, we know the brain 'talks' to our gut and its resident microbes. But *how* does this occur? I asked John this very question. He said at least four critical pathways explain this two-way communication system between our gut microbes and our brains.[13]

The first pathway linking gut microbes to the brain

The first of these pathways is via the vagus nerve – the one I mentioned earlier. This nerve is a vast, meandering structure that reaches far and wide across the body – hence its name, which

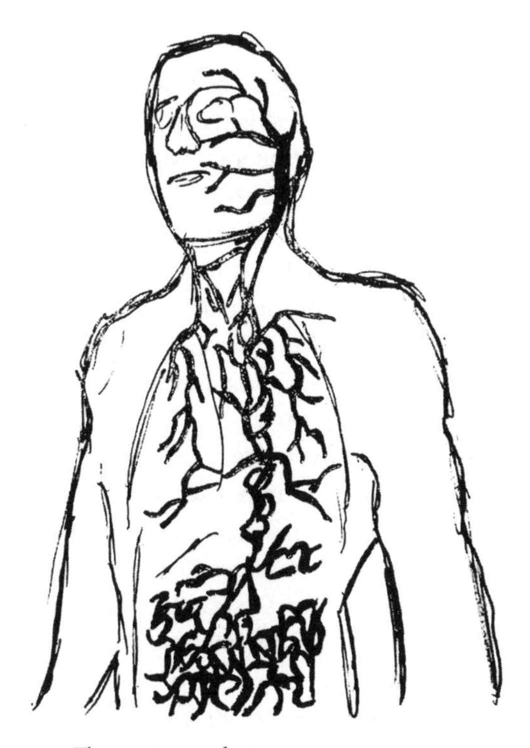

The great meandering vagus nerve.

means 'wandering' in Latin. The words 'vagrant', 'vagabond' and 'vague' all come from the same etymological root.

Historically, the vagus nerve was called the 'pneumogastric nerve' because it connects the lungs (*pneumo*) to the gut (*gastric*). Until 1921, we didn't know how nerves in the body stimulated a particular function or action in a connecting organ, such as the lungs, gut and brain. The possibility that the vagus nerve stimulated the release of chemicals to cause an effect in our organs was put forward in the early twentieth century. Finally, proof of this was obtained in 1921 by a Czech-Canadian pharmacologist named Otto Loewi (1873–1961).[14] Loewi demonstrated that stimulating the vagus nerve did indeed lead to the release of a chemical that, in turn, changed a function in the heart of a frog – specifically, it decelerated the frog's heart rate. This chemical was called acetylcholine, and was the first neurotransmitter (a chemical released by nerve fibres) ever discovered. Loewi and colleague Sir Henry Dale, went on to win the Nobel Prize in Physiology or Medicine in 1936 for this discovery.

Although we still need to do plenty of research to understand how our microbial gut residents stimulate the vagus nerve, researchers think that they interact with special cells in the gut. These cells produce molecules that modify the activity of nerves. They also produce peptides (short chains of amino acids – the building blocks of proteins) that help the vagus nerve detect the microbes with its tiny protruding fingers and transfer information to the central nervous system. Because of this remarkable ability to sense microbes and other compounds, some people refer to the vagus nerve as our 'sixth sense'.

Once the central nervous system receives the information, it is processed and converted into a response. This could lead to a change in our behaviour. The response could also lead to changes in our body's chemicals that influence our health. The vagus nerve comprises 80% *afferent* fibres (strands that send impulses from the bodily organs to the central nervous system) and 20% *efferent* fibres (strands that send impulses from the central nervous system to the organs). Hence, the threads that carry information away from the gut and its resident microbes are abundant in the body.

Picture two cities that contain dense networks of roads connecting all the buildings, and then a large highway connecting the two cities. We can view these two cities as the central nervous system and the gut, and the large highway as the vagus nerve. Now picture a delivery driver collecting packages (or information) from one city and transferring them to another. Much as exchanging packages, goods, foods and hospital supplies between cities is essential for our society to flourish, this exchange of information along the two-way vagus nerve highway is necessary for the health and stability of our body's ecosystem – a state also known as *homeostasis*.

Eubiosis or 'life is good' refers to a diverse, flourishing ecosystem. We often use this term to describe our body's microbial ecosystem when it is in a state that promotes health. Conversely, *dysbiosis* or 'life in distress' refers to our microbial ecosystem when it has relatively few microbial species and is unable to function properly. This often creates the conditions for pathogens to flourish. Our gut microbiome may be in a state of dysbiosis if, for example, we take too many antibiotics or if we regularly consume an unhealthy diet of ultra-processed foods such as crisps, sweets and fast food. Indeed, we need to look after our microbes, so they can look after us. If our gut microbiome is in a state of dysbiosis, this can affect the vagus nerve's ability to function, which in turn can affect our physical and mental health. Moreover, stress can inhibit the activity of the vagus nerve.[15] This can have negative impacts on the gut and its resident microbes, and can promote chronic diseases. It is thought that stress, by affecting the vagus nerve, is involved in gastrointestinal disorders such as inflammatory bowel disease or 'IBD' and irritable bowel syndrome or 'IBS'. These conditions both reflect a state of dysbiosis.

On a lighter note, discovering all the fascinating interactions that occur in this nerve has led to a plethora of quirky scientific publication titles that play on the great gambling city in Nevada. These include: 'Gut microbe to brain signalling: what happens in Vagus', published by John and his team in the journal *Neuron*,[16] and 'What happens in Vagus, no longer stays in Vagus', published in the *Journal of Physiology* … But surprisingly, I found no 'Viva Las Vagus'.

The second pathway linking gut microbes to the brain

The immune system is the second communication pathway that links microbes in the gut to the brain. As described in Chapter 2 the immune system is incredibly complex and interacts with many other systems in the body. Microbes also shape it considerably. John tells me there is direct chit-chat between the immune system and the brain. Our guts harbour the densest concentration of immune cells in the entire body. These immune cells constantly talk to the trillions of microbes residing in our gut. It's like going into the office for a team meeting. You've got different people fulfilling different roles, and you come together so that the organisation (the body) runs smoothly.

This communication can occur directly through physical contact between our microbes and the immune cells, or by exchanging secreted chemical compounds. Our intestinal lining (the meeting room) is where most 'talking' occurs between our gut microbes and our immune system. Bacteria have special molecules called *antigens* on their membranes that allow our immune system to identify them and detect changes in the state of the gut microbial ecosystem. Interactions between microbes and our intestinal immune cells can lead to the release of a flurry of chemicals that carry messages along the spinal cord to the brain. This torrent of information from the gut to the brain (like the torrent of emails that may follow a strategy meeting) is thought to update the body's health status and potentially regulate behaviour accordingly.

Our body's innate immune system (the primary line of defence) includes a group of remarkable cells called *microglia*. Microglia hang out together in huge numbers throughout the brain and spinal cord. Neurons steal much of the limelight for being the cells that do the communicating in our bodies. But microglia make up 10–15% of all brain cells.[17] They have a small body with long thin cellular branches that reach outwards to assess the local environment, keeping a constant eye out for signs of danger.

I've sketched a microglial cell below. This made me pause for a moment and think how amazing it is that everything in life branches out, from tiny microglia and neurons to capillaries, tree roots and river deltas. Not everything in life acts like microglia, however. Microglia provide a security system for our brains and clean up swathes of dead and pathogenic cells. If needed, microglia can change shape and multiply, and produce other cells called *cytokines*. These cytokines make it easier for other immune cells to enter the brain and help control infections, like the T cells mentioned earlier.

Some microglia act like armies of cleaners. They take their broom, mop and bucket and clean up any debris in the brain. Following an infection, microglia send signals to recruit other cells to help heal the damaged tissues. Essentially, microglia are vital in keeping the central nervous system clean, safe and organised. Interactions between microbes and microglia have received a great

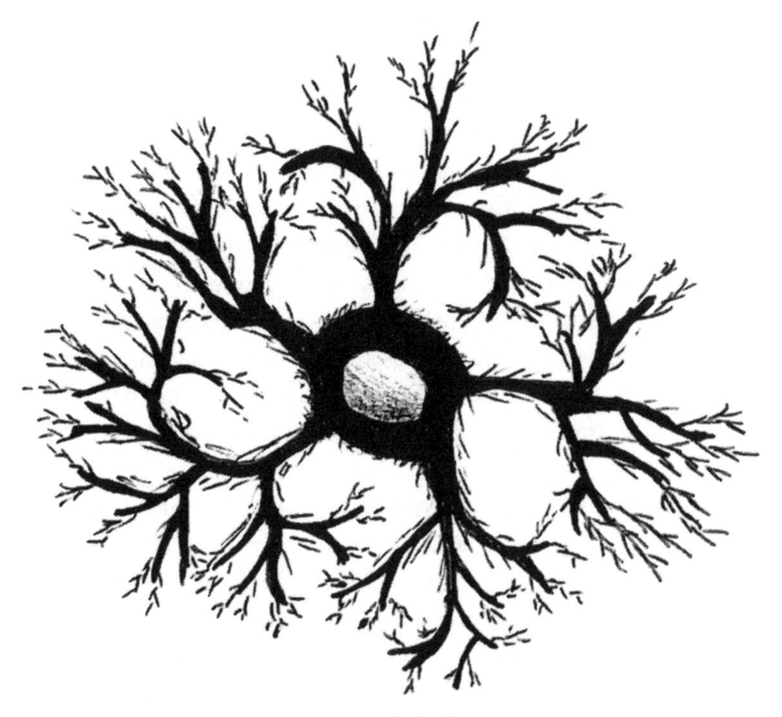

Microglial cell.

deal of attention recently. We now recognise these as critical players in connecting the microbiome, the immune system and the brain.

Researchers recently demonstrated that a diverse gut microbiome is needed to maintain and mature microglial cells, keeping them healthy.[18] Using germ-free mice, researchers discovered that the absence of a diverse gut microbiome could cause defects in microglia. This could compromise the host's immune response to bacterial and viral infections. These microglia defects were reversed when the researchers 'recolonised' the guts of the mice with microbes. Just imagine removing the coach from a top-tier football team: the players could still play football, but the team (or at least its strategy) would now be defective. The football team would probably start to lose games. Bring the coach back, and we restore the strategy and morale. The team begins to win games again.

These studies support the notion that a healthy and diverse gut microbiome is vital for healthy microglia, and thus for a healthy, functional central nervous system. Evidence suggests that gut microbes also help transport immune cells from the body's organs to the brain.[19] This was witnessed when probiotics were given to subjects in a clinical study.

Chitter-chatter between the brain and our gut microbes also occurs through the adaptive immune system. We call this part of our immune system 'specific immunity' because it has a targeted response to a given pathogen. In recent years, scientists have made efforts to unravel the complexities of the adaptive immune system's role in brain function and behaviour. One remarkable study demonstrated that inoculating stressed mice with adaptive immune system cells reduced anxiety and increased social behaviour; importantly, they do this by working synergistically with gut microbes.[20]

Defects in the immune system of germ-free mice can affect their behaviour. But inoculating the mice with probiotic bacteria has reduced these effects. Probiotics can also improve emotional learning in rats. Because the scientific community has a penchant for long, technical names, these *Lactobacillus* bacteria inevitably have one. We call these *facultative anaerobic heterofermentative* species. Don't worry, there's no need to remember this term – let it pass through your eyes and brain and out into a cosmic void. However, it basically means they can grow in the presence and absence of oxygen and undergo fermentation to produce various valuable chemicals. The chemicals likely play a vital role in repairing damaged cells.

Incidentally, *Lactobacillus helveticus* is so named because it is the bacterium used in the production of Swiss cheese (the one with holes). *Lacto* is from the Latin for 'milk', *bacillus* comes from the Latin for 'stick', which describes the bacterium's shape, and the word *helveticus* comes from the Latin for 'Swiss'. The bacterium produces carbon dioxide, making the famous holes in the cheese. These probiotic bacteria are probably helpful not just in making our cheeses, but in maintaining our bodies too.

The third pathway linking gut microbes to the brain

Microbes are like tiny factories, continuously producing chemicals. Some chemicals are made when microbes or other organisms break down food and other substances. One of the most important and well-studied groups of these chemicals is *short-chain fatty acids*. This is probably one to remember, as I mention them a few times in the book.

When scientists sampled the faeces of people with neurological disorders such as Parkinson's disease and anorexia nervosa, they found decreased levels of short-chain fatty acids.[21] Lower numbers of these compounds in the gut are linked to a range of other diseases, from chronic stress to Alzheimer's disease. Gut microbes and their metabolites may also play a role in the behavioural symptoms of autism spectrum disorder. Together, these findings suggest that the chemicals produced by microbes are critical players in gut–brain communication.

Short-chain fatty acids are abundant in our foods, particularly in high-fibre foods such as fruits and vegetables. Because our gut microbes feed on our food, we now refer to the non-digestible fibres as *prebiotics* or 'before life'.

Prebiotics are simply the food for probiotics (health-promoting bacteria). This process has given rise to the term 'you are what your microbes eat'. The levels of short-chain fatty acids in human faeces can be 112 g per kg.[22] The solids can be made up of over 50% bacteria. Therefore, most of your poo is composed of microbes and their chemical by-products.

Several animal studies show links between short-chain fatty acids (particularly *butyrate*) and enhanced learning, memory and sociability, along with decreased depressive-like symptoms. The short-chain fatty acids produced by microbes help our cells chatter away to each other. We call this *cell signalling*, and it's important to the communication between our microbes and our brains.

The bodily systems and processes that keep us alive are entirely reliant upon our cells' ability to communicate with each other. It is imperative that our cells can talk to each other to regulate our heartbeat, stimulate our 'fight or flight' hormones, respond to pathogens and keep our brains healthy.

Imagine you have had an accident on a mountain and need emergency help. Luckily, your friend who is with you has a mobile phone and can call the mountain rescue team. However, to do this, your friend needs (a) to be present, (b) to detect and evaluate your injury, and (c) to have the means of contacting the emergency services. In addition to this, the emergency services will need to request certain information regarding your condition and location and bring you the help you need to survive. This scenario highlights the vital importance of efficient communication between various people to keep you alive. Although it is an imperfect analogy, cell signalling works similarly. It is imperative that our cells can detect when something is wrong in the body and be ready and able to respond through a cascade of cell-to-cell communication. To do this, our cells require two vital elements, along with many others. One is a group of *signalling molecules*. In our mountain-rescue scenario, this is your friend making the emergency call with a phone. The other is a group of *receptors*. Think of this as someone who answers the call on the other end. Other examples include the 'key and lock' analogy, in which a key (the signalling molecule) is needed to open the lock (receptor) to exchange critical information. We know that short-chain fatty acids to be important signalling molecules. Therefore, they play a vital role in cell-to-cell communication from the gut to the brain. They are the keys that unlock the doors to a torrent of information.

Short-chain fatty acids can activate areas of the brain via the vagus nerve (the great meandering cable linking our brain to our gut). Many studies have shown that animals display reduced anxiety-like behaviours when inoculated with short-chain fatty acids.[23] Moreover, the most cutting-edge of research indicates that gut microbe-derived short-chain fatty acids influence the hippocampus in the brain. The hippocampus helps coordinate cognition (thinking) and memory. Although we need more research, particularly in humans, there is a sense of excitement in the scientific community regarding the role of gut microbes in learning and memory. If you're a whizz at remembering dates and learning new things, could this be because you have a diverse gut microbiome that secretes a rich bounty of health-promoting chemicals? We'll explore this concept further in Chapter 7.

The fourth pathway linking gut microbes to the brain

Various glands secrete hormones throughout the body. As alluded to earlier, our microbial friends also produce and secrete hormones inside our bodies, respond to our hormones and regulate the expression of our hormones. Because hormones play an essential role in our brain and behaviour, such as priming our response to stressful situations, scientists think our gut microbes may affect our brain and behaviour through hormones.

The *hypothalamic-pituitary-adrenal axis*, commonly known as the HPA axis, is vital in this relationship. The HPA axis helps maintain a stable environment in our bodies. It is one of the main routes of communication between our gut microbes and our brain. When we face a stressful situation, perhaps one in which we need quick decisions and escape routes, our endocrine system releases a cascade of hormones to kick-start the body's response. Firstly, the hypothalamus, located in the centre of our brain, sends a message to the pituitary gland, asking it to secrete a hormone. This hormone acts upon the adrenal cortex, located much further down the body, sitting snugly above the kidneys. Our adrenal cortex then releases *glucocorticoids* – a class of steroid hormones. Increased levels of glucocorticoids in the bloodstream promote the mobilisation of amino acids (the building blocks of proteins) and the breakdown of fats to maintain sufficient glucose levels in the bloodstream. This additional energy allows us to mount an appropriate stress response.

This is fantastic if we need to think quickly and escape an immediate threat – for example, running away from a predator. That is partly why the system evolved. However, if we respond to stress inappropriately and glucocorticoids are secreted too regularly, for instance when we're stressed by standing in a long queue, being late for a meeting or being stuck in traffic, then these steroids can start to cause severe damage to our bodies because of the build-up of atherosclerotic plaques in our vessels.

In his book *Why Zebras Don't Get Ulcers*,[24] Stanford neuroscientist Professor Robert Sapolsky describes the process and impacts of inappropriate stress responses. This book profoundly changed how I viewed and responded to everyday 'modern' situations that trigger the fight or flight stress response that our ancestors evolved to escape predators on the great African plains hundreds of thousands of years ago.

Gut microbes are linked to the stress response system in animals. Mice with dysbiosis (a poor microbiome) show an overreactive stress response, suggesting that microbes help regulate stress.[25] Interestingly, the stress response system also influences gut microbes. Scientists have observed considerable changes in the make-up of the gut microbiome when the stress response system is altered. Therefore, much like our old buddy the vagus nerve, it seems the HPA axis–gut microbiome system is a two-way street.

I was fully aware there would be many to choose from, but I asked John what he considered the most exciting finding in his research career. He said one of them was discovering an increase in myelin in the brains of mice when their gut microbes were absent. Myelin is an insulating layer – also known as a 'sheath' – that wraps around our nerve cells. The myelin sheath allows electrical impulses to transmit quickly and is vital to the normal functioning of the nervous system. Too much myelin, or inappropriate myelination (the process whereby myelin builds up around our nerves), and strange and dangerous things can happen to our central nervous system. There are also situations where the body's immune system goes awry and can end up attacking the myelin sheath. This is what we refer to as an autoimmune disease, and a prime example of this is multiple sclerosis (MS). Multiple sclerosis is a debilitating disease, and symptoms can include severe issues with the body's vision and coordination.

Our nerves are like the electrical cable to a lamp, sending impulses from an electrical source (the plug socket) to the lightbulb. The myelin sheath on a neuron is like the plastic cover around the electrical cable. If the cable's cover deteriorates, the cable could short out from being unprotected, and the light will flicker or fail.

It has been shown that people with MS are more likely to have dysbiotic gut microbiomes, including a reduced number of microbial species, than control groups (groups of people who did not have MS).[26]

We need additional research to fully understand the direction of the relationships in these studies. For instance, does MS lead to dysbiosis, or does dysbiosis trigger the onset of the disease?

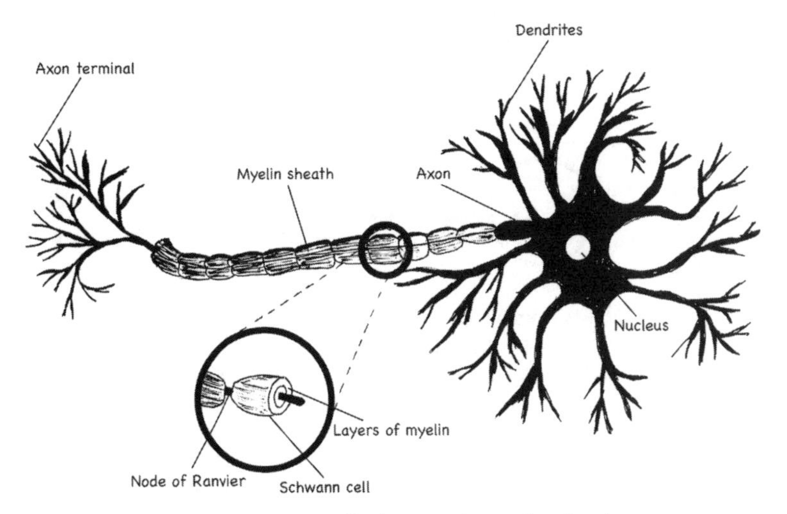

A neuron (nerve cell), showing the myelin sheath.

Nonetheless, the collective evidence offers hope. Continuing to study gut microbe interactions provides the hope of understanding more about how MS works – and, dare I say, with crossed fingers and toes, how it could potentially be alleviated.

One other recent development in microbiome research that John thought fascinating was the 'mouse transplant studies',[27] a series of experiments in which microbes from sick, old and obese mice were transplanted into healthy, young and slim mice. Following the microbiome transplants, the healthy, young and slim mice became sick, exhibited ageing disorders (e.g. impaired memory) and became obese. This is some of the most compelling research to date showing that gut microbes have fundamental roles in the onset of diseases.

Striding a step further, a human twin cohort study (which allowed researchers to control for genetic variation) demonstrated that when scientists transplanted the gut microbes of obese human twins into germ-free mice, the mice gained significantly more fat than the mice inoculated with gut microbes from lean human twins. This clearly shows that gut microbes have a crucial role in developing obesity in mice, and potentially in humans too.

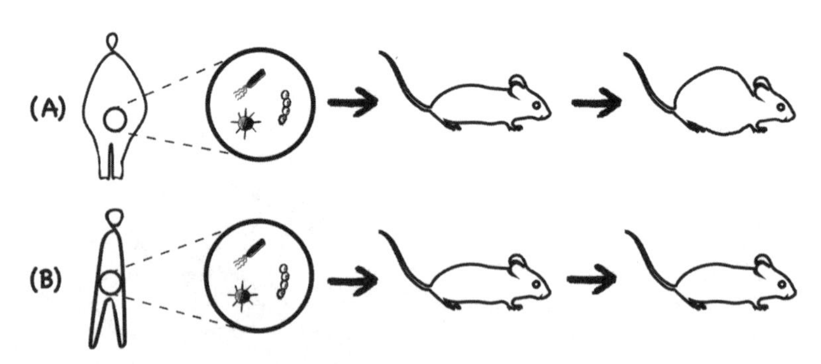

(A) When microbes from obese twins were transplanted into germ-free mice, the mice became obese.
(B) When microbes from lean twins were transplanted into germ-free mice, the mice remained lean.

I asked John, 'What are some of the real-life applications of your work, and what are the future research directions?'

Firstly, John was keen to point out that none of this fantastic research would be possible without the exquisitely talented group of researchers and technicians around him. John leads a team back in the lab in Ireland, including neuroscientists and microbiome clinicians, data crunchers called 'bioinformaticians', laboratory technicians and managers, and a large cohort of PhD researchers. It is a finely tuned symphony orchestra of scientists; each person is essential to the ensemble. Before going into detail, John summarises his answer to my question by saying: 'Our current and future aim is to understand how we can mind our microbes to look after our minds.'

This pithy comment underscores what some consider a *psychobiotic revolution*. That is, we are on the cusp of understanding how we can look after our internal microbial ecosystem by providing the right conditions and the proper nutrients to support the right microbes, to ensure we have healthy bodies and healthy minds. After all, we're not in this alone. One fact emphasised throughout this book is that each of us is a system comprising trillions of other organisms that work symbiotically to sustain our health as a whole. I remember someone once said to me: 'Humans are literally incapable of doing anything by themselves.' Indeed, we are entirely reliant upon our invisible friends. John and his team strive to enhance human health and well-being by understanding how best to support our microbes so that they can support us.

Finally, John emphasises that microbiome research is not a niche phenomenon. Microbes are fundamental to all life on Earth, and they influence all aspects of society – from ecology and medicine to philosophy and sustainability. If we can work together collaboratively, across disciplines and communities, we can help microbes make the world a better place for humans and the rest of nature.

I squeeze in a final question: 'What can microbes teach us about the world?'

'Microbes can teach us to be humble,' John replies. Given that we would not exist without them, I agree with him. After all, we're living in a microbial world.

CHAPTER 6

The Lovebug Effect

'Nature isn't just beautiful, intriguing, and awesome, it is our life-support system.'
—Lucy Jones

Let nature be thy medicine. While I was studying for my PhD, I lived in the Peak District National Park in north-central England. My wife and I regularly ventured into the wooded dales of the charmingly named Hope Valley region. We would cycle around for a couple of hours, taking in the countryside's crisp breeze and diverse scents. We were cheered on by baaing sheep and bedazzled by the barn owl and the kestrels' majestic hunting displays; these trips were our inspiration, our fortitude, our medicine. For me, cycling in the hills helped to clear my noisy mind. I found this ridding of mental clutter would often help when it came to conceptualising research ideas. Most ideas are reconfigurations and extrapolations from other people's past and present thoughts.

In late 2019, I started to think about how the microbial denizens in our bodies might influence our decisions to spend time in particular environments. In my mind, I had a jigsaw puzzle to complete. The pieces to my jigsaw were in the form of Graham Rook's ideas on human–microbe co-evolution, John Cryan's studies on the microbiota–gut–brain axis, E.O. Wilson's work on biophilia (which I'll define shortly), and the more recent research on nature connectedness – one's emotional, cognitive and experiential connection with the rest of the natural world. I could see clear links between these phenomena, and joining the pieces together culminated in publishing a paper in 2020 with co-developer and author Dr Martin Breed, called 'The Lovebug Effect: Is the human biophilic drive influenced by interactions between the host, the environment, and the microbiome?'[1]

Lovebug Effect translates to 'microbio-philia', from 'bug' – a colloquial term for a microbe, and '*philia*' – a Greek word for love or attraction. The Lovebug Effect hypothesises that the invisible biodiversity inside and around us could influence our desire to affiliate with natural environments. The hypothesis builds upon research and theory from microbiome science, environmental psychology and evolutionary biology. We argue that co-evolution occurs between humans and microbes, and by altering our brain chemistry, microbes may be able to manipulate our behaviour – for example, to influence the food choices we make, or importantly for this paper, our decisions to visit specific environments that may confer a benefit to our body's microbes.

As discussed in the previous chapter, the research by John Cryan and his team highlights the presence of a microbiota–gut–brain axis – the complex two-way communication highway between the gut and the brain. By producing an array of chemicals in the body, we think that microbes residing in the gut essentially 'hijack' this communication highway. This could have important implications for human behaviour. One manifestation of this could be a desire to spend time in biodiverse environments (those containing lots of plants, animals and microbes). This action would likely benefit both the human host and its microbial ecosystem. It's like if you're running out of food at home, and someone in your household asks you to do some food shopping – they recognise that the health of the residents in the household will suffer if they don't have nourishing foods. So, you head down to the supermarket and fill your trolley. The foods will benefit both you and your partner, and any other resident. Like the person in your household, your microbes could be talking to your brain and subconsciously convincing you to visit a place with nourishing products (e.g. diverse microbes and other compounds in your local woodland) that will benefit both you and your microbial residents.

The concept has certain parallels with the *biophilia hypothesis*. American evolutionary biologist Edward O. Wilson first proposed this in 1984.[2] Wilson described biophilia as the 'urge to affiliate with other forms of life', or the 'connections that human beings subconsciously seek with the rest of life'. The Lovebug Effect builds on this hypothesis and provides a biological mechanism for our evolutionary desire to affiliate with biodiverse environments.

We explored the possibility that humans could be classed as 'metaorganisms', hosting trillions of microbes that work in a mutualistic relationship with the human body, mind and environment to form a flourishing, walking, talking ecosystem. There is good reason for humans and microbes to 'work together' to sustain the health and stability of this dynamic conglomerate we call the human body, because all parties benefit from doing so. Part of the Lovebug Effect involves microbes influencing our decisions to spend time in more biodiverse environments, immersed in dense clouds of invisible friends and a vast array of organic chemicals that confer vital health benefits. We are exposed to far more microbial species by visiting natural environments than in places where nature is kept out.

Our paper also discusses other pathways to this microbe-driven nature affinity – for instance routes that do not involve direct manipulation of our behaviour by microbes. These include potential changes in inherited traits (behaviours) shaped by close-knit relationships between humans and microbes across generations, particularly those that maximise survival and reproduction. This is much like my example in the Introduction, whereby *Streptomyces* co-evolved with springtails, luring them in with attractive odours. The springtail benefits from a nutritious food source. The *Streptomyces* bacterium benefits from being dispersed far and wide, and thus increases its chances of securing its future on the planet. This is a prime example of co-evolution.

How the natural environment benefits us and our microbial friends

A recent study I was involved in showed that urban habitats with higher vegetation diversity (e.g. those with grasses, shrubs and trees) harbour a much broader range of bacteria in the soil and the air compared to environments with low levels of vegetation or 'monocultures', such as sports fields.[3]

For this study, which was part of my PhD, I spent three months in Australia. Once again, the study was co-designed with Dr Martin Breed at Flinders University. For the first week, I spent the humid days sawing wood and tightening screws to build microbiome sampling stations. An Eastern blue-tongued lizard and a couple of rosella parakeets regularly joined me; the parakeets came to escape the heat and dip their vivid multi-coloured feathers in the water bath.

Once I built the sampling stations, I ventured out into the Parklands of Adelaide with my Ecuadorian friend and colleague, Chris Cando-Dumancela. We spent several days collecting the airborne and soil-dwelling bacteria using Petri dishes and tubes, and froze them before heading to the lab. In the lab, we would extract the DNA from the bacteria. Each bacterium's DNA sequence was then deciphered using a special sequencing machine. This allowed us to see which bacteria were present in the samples. By doing this, we could witness how diverse the microbial communities in the environment were and identify which bacteria might be helpful or pathogenic.

As mentioned earlier, environments with more diverse vegetation tend to harbour a greater diversity of bacteria. And if you recall from Chapter 2, a diverse microbiome (i.e. one with more species) is considered better for our health. When a greater diversity of invisible friends surrounds us, more species become available for different bodily functions, and opportunistic pathogens can be more easily outcompeted. Just like a group of friendly children might stick together to oust the school bully.

Our study found a far greater diversity of friendly bacteria and far fewer pathogens in habitats with shrubs and trees, compared to less natural 'monoculture' habitats such as sports fields or amenity grasslands. We also found that species diversity was much higher in the samples collected closer to the ground, compared with those collected higher in the air column.

We had predicted this bacterial diversity lower down, because the soil is one of the most biodiverse habitats on Earth, with as many bacteria in a thimbleful as humans roaming the whole planet.

Soil and vegetation are primary sources of the bacteria floating around in the air and entering our bodies. Therefore, it's beneficial to spend time engaging with these biodiverse environments. We know that at a population level, our health suffers when we remove the biodiversity around us. In doing so we are, in essence, starving our internal microbial ecosystems of the life-supporting invisible friends needed to ensure our microbiomes can flourish, particularly in early life.

Other research has demonstrated that when more trees are present in the urban environment, the abundance of pathogenic bacteria (those relatively few species that cause human maladies) decreases.[4] Venturing out into the wild thus has enormous potential to recharge our internal ecosystems with diverse, health-promoting microbes whilst reducing the likelihood that we'll pick up the nasty ones.

Scientists in Adelaide have shown that spending half an hour in a green space can significantly change the adult human nasal and skin microbiomes.[5] Another study on mice showed that trace levels (i.e. tiny amounts) of biodiverse soil dust significantly altered the gut microbiome of the mice.[6] Moreover, certain species that produce short-chain fatty acids were present, which correlated with reduced anxiety-like behaviours in the mice.

Given the benefits of exposure to diverse microbial communities from the environment, it seems plausible that we have co-evolved mechanisms to draw us closer to such environments as an evolutionary default. And as we humans have removed our ancestral natural habitats on a vast scale since the Industrial Revolution, rapid microbial evolution could have given rise to such a trait – also known as 'biophilic drive'.

When our gut ecosystem lacks species and relationships that sustain homeostasis, an evolutionary response might be to seek out these core invisible friends. However, the extremities of current Western-style diets, the lack of biodiversity, the overuse of antibiotics and increased exposure to industrial pollutants could all inhibit the Lovebug Effect by causing a range of non-infectious maladies such as diabetes and inflammatory bowel disease. These diseases could lead to serious microbiota–gut–brain connection issues. No surprise, then, that we're witnessing a torrent of these diseases worldwide, which are deeply linked to our changing and increasingly urbanised lifestyles.

The extended phenotype and mind-altering microbes

The notion of behavioural manipulation at the metaphorical hand of a microbe is by no means a novel concept. The central theorem of the *extended phenotype* proposed by English evolutionary biologist Richard Dawkins suggests that the genes that influence host behaviour tend to be maximised because of the behaviour itself, regardless of whether the genes are of host origin (i.e. they could be of microbial origin instead).[7]

We can look at the classic example of host behavioural manipulation by the small but mighty protozoan *Toxoplasma gondii*, which from this point onwards I'll refer to simply as '*Toxo*' – its colloquial name. *Toxo* is a eukaryotic microbe, which means it has a nucleus enclosed in a tiny envelope, unlike prokaryotic microbes such as bacteria and archaea in which the DNA floats around on tiny lilos. *Toxo* can only complete the reproductive stage of its lifecycle in the intestines of species in the Felidae family – also known as cats. Therefore, cats are the *definitive hosts*.

Cats eventually shed *Toxo* cells in their faeces. At this point, the cells undergo sporulation and become infective. Notably, another group of animals called *intermediate hosts*, including rodents and birds, come along and gobble up the infective cells left in the cat's poo or nearby soil. These intermediate hosts are typical prey items on the menu for cats. It so happens that *Toxo*'s survival depends on the cat becoming infected by feeding on the infected prey. Scientists think this survival pressure led to *Toxo* evolving special powers to manipulate the behaviour of the intermediate host – that is, the rodents. Incredibly, this manipulation results in the rodent losing its fear of cats; you may consider it a 'fatal attraction'.

I can picture the mice or rats jumping up and down, screaming, 'Yoo-hoo, I'm here, come and get me!', whilst their non-infected mouse friends look on from a distance saying, 'Crikey, Steve's lost the plot!' This bewitching example of evolution and behavioural

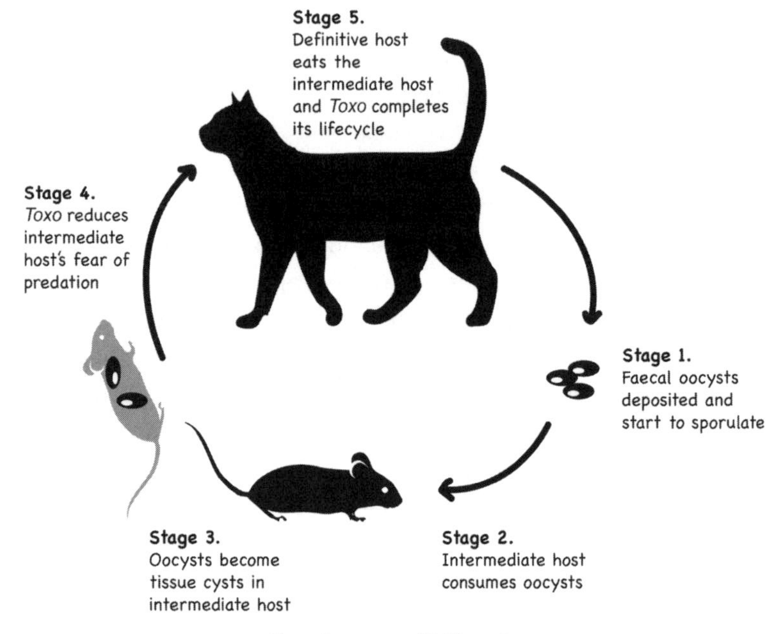

Toxoplasma gondii lifecycle.

manipulation increases the transmission of parasite genes into future generations.

Although some of the underlying mechanisms are still unclear, we think that *Toxo* stimulates testosterone production to influence the amygdala in the brain, which then leads to the loss of innate aversion to predators.

The amygdala is the part of the brain responsible for processing fearful or threatening stimuli, so manipulating this organ can be deadly. However, there are exceptions to this rule. The American rock climber Alex Honnold has an under-reactive amygdala, which means his fear aversion is exceptionally low. Alex mostly climbs in a 'free solo' style,[8] tackling giant rock faces over 2,000 metres high without ropes and safety equipment. To many people, this probably sounds like Alex has a death wish. But he is scrupulous in his planning, ensuring he knows every nook and cranny using ropes before going ropeless. The one thing that he does not need is fear, as this would lead to mistakes. Fear would be a deadly trait for Alex, so in this unusual case, an under-reactive amygdala could be a lifesaver – unlike for the *Toxo*-laden mice, whose fear-losing amygdala is deadly!

It is worth noting that it's not just the behaviour of animals that symbiotic microbes can influence. A recent study demonstrated that a bacterium living in flowers, called *Acinetobacter*, acquires a nutritious food source in the form of pollen by inducing pollen production and maturation.[9] That is, by 'hijacking' the germination mechanisms of pollen, the bacterium alters the flower's behaviour to benefit the bacterium's fitness, meaning the flower might germinate at an unexpected time.

The Lovebug Effect as a recent phenomenon

Unlike evolutionary processes in animals, *microbial* evolution can be remarkably rapid. This occurs due to the ability of many microbes to replicate rapidly and even transfer their genes into other organisms in real time and without mating. As mentioned earlier, this process is known as horizontal gene transfer, and there are three mechanisms by which this can occur in bacteria.

The three horizontal gene transfer mechanisms are: (1) *Transformation*, whereby a DNA fragment from a dead and degraded bacterium enters a living recipient bacterium and is exchanged for a piece of DNA of the recipient; (2) *Transduction*, which involves the transfer of a DNA fragment from one living bacterium to another living bacterium by the spider-like bacteriophage viruses, which carry the DNA from one and inject it into another; and, last but certainly not least – in fact, this is the most likely mechanism – (3) *Conjugation*. This involves the transfer of DNA from a living donor bacterium to a living recipient bacterium by cell-to-cell contact.

Imagine if you could receive strands of DNA from another person that could be inserted into your genome just by giving them a big hug. This is what horizontal gene transfer is like, just on a microscopic level.

Horizontal gene transfer can be excellent for the bacteria – allowing them to rapidly adapt – but it can be rather menacing in

Conjugation

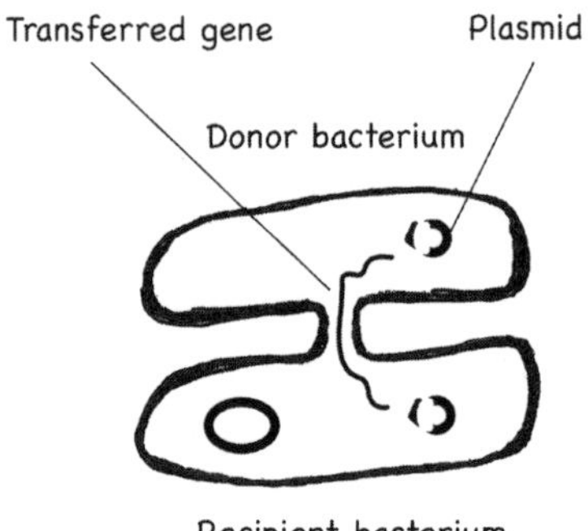

Transduction

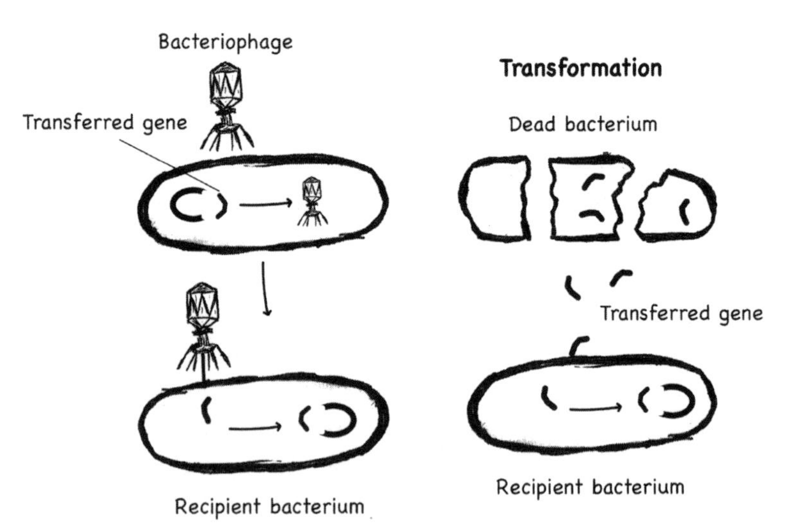

Horizontal gene transfer.

a human context. For instance, it can provide a platform for the spreading and persistence of antibiotic-resistance genes. In turn, this makes it much harder for us to treat and control bacterial diseases.

Due to this microbial ability to rapidly evolve, if the selection pressures associated with the proposed Lovebug Effect developed following the era of the Industrial Revolution, rapid global biodiversity loss, sterilisation of microbes en masse, and our loss of connection with nature, then microbial mutations could conceivably still have occurred in this relatively short time frame. This would ensure the stability and meet the demands of the human's internal microbial ecosystem, and proliferate the microbes' genes.

The Lovebug Effect could add to our understanding of psychological frameworks often used to investigate our affinity towards and connection with the rest of nature, such as the biophilia hypothesis and nature connectedness. The human gut microbiome remains relatively stable after the first few years of life. Therefore, environmental microbes likely have their most significant impact on humans before and during the initial weaning period. In our Lovebug Effect paper, we ponder whether parents with a high emotional and cognitive connection with nature (a high level of nature connectedness) might influence their child's microbiome development through the increased likelihood of spending more time in nature with their child. This scenario could have important implications for the child's health and development in later life.

As mentioned, a dysbiotic human microbiome (which is analogous to a degraded ecosystem, such as a felled woodland or bleached coral reef), created for example through the overuse of antibiotics along with a poor diet, could perturb the Lovebug Effect. This could conceivably lead to continued physiological or behavioural impacts that reduce a person's emotional, cognitive and experiential connection with nature.

Indeed, people with higher levels of nature connectedness are more likely to exhibit higher levels of well-being. And well-being is inversely associated with depression. Moreover, depression has been linked to a dysbiotic microbiome.[10] This is a tantalising and potentially impactful relationship to investigate. One question that is worth asking is: might spending more time in nature boost our microbiomes, and in turn reduce symptoms of depression?

If proven accurate, the Lovebug Effect could have important implications for our understanding of exposure to natural environments for health and well-being. It could also contribute to an ecologically resilient future in which exists widespread reciprocity with the natural world.

CHAPTER 7

The Holobiont Blindspot

'In the Mind there is no absolute, or free will, but the Mind is determined to will this or that by a cause that is also determined by another, and this again by another, and so to infinity.'
—Baruch Spinoza

Have you ever thought about cognitive biases? These are systematic errors in judgement, also known as 'cognitive blindspots'. As human beings, we're prone to making these errors every day. They can have real implications for our relationships, empathy and work-life.

Let's look at some examples of cognitive biases already identified. Researchers in the field of psychology have classified oodles of these blindspots. One of the most common is *confirmation bias*. This describes the tendency to focus on information that only supports our existing preconceptions. Hence, confirmation bias often involves ignoring information that challenges our assumptions and beliefs. Social media platforms use sneaky algorithms and advertising tools to exploit this bias and attract users to new products. However, we can see even darker aspects in the disinformation campaigns that exploit confirmation bias by tailoring messages to people who already believe in them. This can eventually create echo chambers, which can be destructive by reducing diversity of thought. Echo chambers are closed systems of communication that amplify preconceptions by repetition. They are insulated from rebuttal and are often politically and socially polarising.

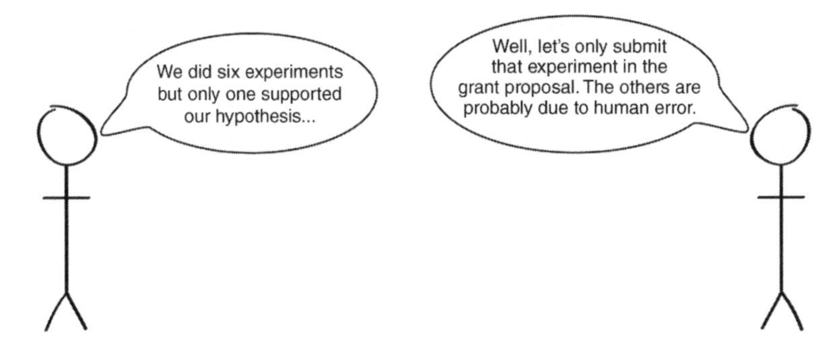

Confirmation bias at work.

Then there's the *bandwagon effect.* This is a phenomenon whereby the uptake of beliefs or ideas depends on how many others have adopted them. As additional people believe in these ideas or opinions, others will 'jump on the bandwagon' regardless of the supporting evidence. We refer to this tendency of people to align their beliefs and actions with those of a group to as 'herd mentality'. By falling for this bias, which is motivated by popularity, there is a risk we effectively clone superficial thoughts and interventions rather than foster independent critical thinking.

Another example of a cognitive blindspot is the *false consensus bias*. This causes us to see our own behavioural choices and judgements as ordinary and overestimate how many people agree with our beliefs, preferences, values or behaviours. It's relatively easy to see how this could lead to relationship issues, whereby one partner

The bandwagon effect – 'everyone else is doing it'.

assumes the other agrees with their choices or views. I notice this blindspot often in ecology and conservation – and I sometimes catch myself falling for it too. For example, because conserving and restoring biodiversity is one of my passions, I am emotionally invested in this behaviour. However, the false consensus bias plays out as I sometimes – and wrongly – assume everyone else does or should feel the same way as me! This blindspot has a calamitous role in conversations and partnerships in all walks of life. It's completely natural to overestimate the degree to which other people agree with us. However, there are strategies to avoid or reduce the frequency of the false consensus bias. One thing you can do is reflect on your views and put yourself in others' shoes. Another is to sit down with an honest and perhaps cynical friend and describe your project or beliefs. Ask them to write down all the assumptions you're making as you do so. Afterwards, you can discuss the evidence and roots of your assumptions. This will help to identify blindspots.

A final example, but by no means the final blindspot, is the *availability heuristic.* This is the phenomenon of people tending to overestimate the importance of information available to them. It is a mental shortcut – the basic definition of a heuristic – that relies on immediate examples springing to mind when evaluating a particular concept or decision. For instance, if you knew someone who lived to a great age but who also had an unhealthy lifestyle, you may think that being healthy isn't a good predictor of longevity

False consensus bias – why couples argue over what to watch on TV.

The availability heuristic could be bad for your health.

after all. Or, if the village you live in has cold weather consistently for a few years, the availability heuristic may manifest as a belief that climate change or warming, more broadly, does not exist.

Overcoming the availability heuristic often requires an active shift in perspective. I was reminded of this recently when I moved into a new home. When I looked out of the window, I noticed the name of the square outside, 'Laughter House Square'. *How jolly*, I thought to myself. Then I walked into the next room and looked out of the window again. It turns out a tree was covering the first letter of the name, which happened to be an 'S'. This was a trivial but amusing reminder that things aren't always as they first seem.

We often need to shift our perspective to address biases.

The notion that we can miss important information from seeing only what is right in front of us serves well to introduce the *holobiont blindspot*. Firstly, let's define a 'holobiont'.

'Holobiont' is a word popularised by the late Professor Lynn Margulis in the early 1990s.[1] The term describes a host organism such as an animal or plant and its associated microbes or 'symbionts'. The collective genome of the host and the symbionts is the *hologenome*. As discussed, a rapidly growing body of evidence suggests that the microbes in our guts can influence our behaviour, mood and decision-making via the microbiota–gut–brain axis.

Cognitive biases can manifest when we attribute human-like behaviours or traits to non-human animals, or fail to understand the true reason for an animal's behaviour because it may seem unhuman-like. For example, assuming a cat enjoys a purple fluffy toy because *we* find it attractive, when the cat would probably be just as happy playing with a less attractive (to us) screwed-up piece of cardboard. Or when a dog barks due to an instinctive reaction to certain stimuli but we may find it annoying because a human wouldn't normally make a loud noise out of the blue. Treating holobionts as individual subjects divorced from any cognitive influence of microbial interactions could be viewed in the same manner. In other words, failing to recognise the role of host–microbiome interactions in behaviours and decision-making underpins the holobiont blindspot.

I recall sitting in the University Arms pub in Sheffield, pondering cognitive blindspots. I had recently read two books that most certainly shaped my thinking. The first was *Beyond the Brain: How the Body and Environment Shape Animal and Human Minds* by Professor Louise Barrett.[2] The second was *Thinking, Fast and Slow* by 2002 Nobel Prize winner Professor Daniel Kahneman.[3]

In her book, Barrett presents a thought-provoking narrative about how we understand and misunderstand animal behaviour. Instead of focusing purely on the brain as the sole determinant of behaviour, Barrett provides a refreshing exploration of behaviour resulting from a set of complex interactions between the brain, the body and the environment. This triggered a thought. If we attribute certain behaviours to our personalities, but our resident microbial symbionts drive them, wouldn't this be a critical cognitive blindspot? And if we're studying animals or plants and notice a variation in behaviour, what if this could be partially attributed to interactions between the host and its microbiome?

In Daniel Kahneman's book, he describes how two systems of thought shape our judgement. Kahneman posits that System 1 thinking involves decisions based on intuition and associative memory, and System 2 involves decisions made by slow, reflective thinking. An example of System 1 thinking is intuitively knowing that one car is more distant than another, whereas you would require System 2 thinking to consciously calculate the distance between the cars and yourself.

Kahneman shows us how System 1 thinking can lead individuals to make snap judgements and erroneous decisions based on heuristics (mental shortcuts) and intuition. Could mind-influencing microbes also be considered in the dimension of System 1 thinking – that is, decisions driven by perception, intuition and associative memory? If this is the case, there could be important ramifications for our understanding of the concepts of perception and mental impulses.

I carried on sipping my coffee while gazing through the window at the industrious bumblebees on the cotoneaster bushes. I then decided to jot these ideas down. I started to build a bridge between the concepts and thought they might make a good 'perspective'-style publication. I ran the ideas past Dr Ross Cameron, my PhD supervisor at the time. The result was a paper titled 'The holobiont blindspot: relating host–microbiome interactions to cognitive biases and the concept of the "Umwelt"'.[4]

Because we can view most multicellular organisms (e.g. plants and animals) as holobionts, we may be missing a thorough explanation for different behaviours if, indeed, we fail to recognise the potential role of the host's microbiome in a given behaviour. Moreover, the microbiome could be an essential component of the system that gives rise to an organism's perceptual world. A fantastic paper by Bueno-Guerra (2018) proposed a broadening of our understanding of an organism's perception and actions based on the under-appreciated role of social dynamics.[5] Bueno-Guerra studied the behaviour of chimpanzees, and found that if researchers projected the human behaviour of 'cooperative bonding' on their chimpanzee subjects, the result was delusive generalisations concerning social behaviours. This included inconsistent results in task-solving scenarios that required cooperation. It suggested that evolutionary behavioural pathways might not be identical in other species with different social dynamics. In other words, some fundamental human behaviours are not observed in closely related species.

This inspired our paper. We suggested that a host's microbiome – which can differ considerably between individuals – might affect behavioural dynamics and lead to delusive observations. For example, we could partially attribute personality traits or even daily variation in behaviours to changes in our inner microbial ecosystem.

A recent animal study demonstrated that gut bacteria could override host sensory decisions by mimicking the functions of certain molecules.[6] In this study, a commensal gut bacterium called *Providencia* produced a chemical called tyramine. This chemical acts on the host's olfactory system – the part of the body responsible for processing smells – and regulates the host's aversive response to certain odours. Scientists believe that this process drives mutually advantageous food decisions, whereby the host is essentially manipulated into selecting food that benefits both the host and the bacterium.

Imagine if this is the case in humans – our gut microbes hijacking our olfactory system to make us crave a particular food or drink. Perhaps that's why I love red wine and chocolate so much. There is evidence to suggest that our gut microbes may indeed appreciate foods and drinks that contain high levels of polyphenols called *flavonoids*. It so happens that berries, red wine and dark chocolate all have high levels of flavonoids. Importantly, flavonoids are incredibly good for our health, and they are linked to lower blood pressure and reduced inflammation in our bodies. Could our microbes influence our choice of certain foods with these dual benefits? It's a tantalising thought, and researchers hope to investigate this in humans in the coming years.

The *Providencia* research represents just one of several recent animal studies demonstrating that resident bacteria can regulate host behaviour. Our paper discusses how two groups of bacteria can work synergistically to manipulate host feeding decisions in fruit flies.[7] Other studies show that bacteria can manipulate host behaviour through olfactory (odour) pathways. For example, individuals can be attracted to chemicals emitted by the fungus *Saccharomyces cerevisiae* and the bacterium *Lactobacillus plantarum* but repelled by chemicals emitted by *Acetobacter malorum*. Microbes can also trigger olfactory responses in animals such as zebrafish and mice, thereby affecting sociability and breeding.

The precise mechanisms need to be unravelled. Still, one thing is clear: the microbiome can influence host perception of stimuli via sensory routes (e.g. by playing with the host's 'smelling' system) and behaviours or decision-making in animals. This concept could have important implications for our fundamental understanding of 'thinking' and behaviours.

Just as cognitive biases can manifest by attributing human-centric behaviours to non-human animals (also known as anthropomorphism), treating holobionts as individuals divorced from any cognitive influence by their microbial residents could also be viewed as a cognitive bias – that is, the holobiont blindspot.

We should also recognise that plants and even microbes can be holobionts. Merlin Sheldrake elegantly articulated this in a recent popular science book, *Entangled Life*. Below, I shall paraphrase a passage from Sheldrake's book to highlight the point.

Sheldrake went to a microbiology conference in South America. Someone stood up to talk about tropical plants that produced a group of compounds in their leaves. Scientists had previously thought that the compounds were a defining feature of the group of plants. However, it transpired that the fungi residing in the plant's leaves produced the compounds. This meant we had to revisit our idea of the plants – an entirely different kingdom of life produced one of their defining features. Or so it seemed. But then another researcher interjected. They suggested that the fungi inside the plant's leaves might not be the source of the compounds; it could be the bacteria living inside the fungi. Our idea of the individual had deepened: the image of a 'Russian doll' springs to mind.[8]

When I was sitting in the pub in Sheffield conceiving the holobiont blindspot idea, I primarily had the third-person perspective in mind – that is, a researcher studying a holobiont. But we can also consider it from the first-person view. Indeed, we can view it in the realm of Kahneman's System 1 'fast and automatic thinking'. A potential cognitive bias could occur if we assume that a behaviour of our own such as an aversion to a smell is purely the result of human intuition and associative memory, when it could result from our microbial puppet masters (e.g. via olfactory manipulation).

We know that the olfactory system is a critical player in social behaviour in humans. Odours can significantly influence appetite, memory recall, purchasing behaviour and sexual arousal. Hence, the holobiont blindspot could have significant social ramifications.

Here's a quick thought experiment to illustrate this point:

1. Imagine that environmental stimuli such as pollution exposure, dietary changes and antibiotics lead to changes in our microbiome.
2. This could then lead to changes in our perception of odours or sex pheromones via the olfactory system – at the hand of our microbial residents.
3. This process alters our preferences, such as changes in how we perceive and prefer different human odours, known to be significant 'attractants'.
4. Could a person become less attracted to another person because of this hypothetical cascade of events?
5. Theoretically, this could have significant social ramifications, leading to relationship issues.

Indeed, scientists have demonstrated that the microbiome can influence mating preferences in animals. One study took a population of *Drosophila* flies and divided them into two subpopulations.[9] The researchers reared one subpopulation on a molasses nutrient and the other on a starch nutrient. The two groups were then allowed to mix. Astoundingly, the flies raised on molasses chose to mate with other flies reared on molasses. And the flies raised on starch? You guessed it: they preferred to mate with other flies raised on starch. The researchers then treated the flies with antibiotics, severely changing their microbiomes. This action abolished the mating preference in the flies. This suggested that the microbiome was responsible for the mating preferences.

The researchers decided they needed more support for their findings. They re-inoculated the flies with microbes from the molasses and the starch media. Incredibly, the mating preferences were restored. We think that microbes change the levels of sex hormones in the hosts to influence mating behaviour.

I find the idea that microbes could influence our moods and mental health fascinating. In the case of depression, people often do not know why they feel depressed. If you feel depressed but have no idea why, it is impossible to articulate it, which is problematic in itself. Could this 'unknown reason' sometimes be a holobiont blindspot? It seems likely that there is a microbial link to depression, at least in some cases. Therefore, going back to John Cryan's psychobiotic revolution, developing interventions around microbial therapeutics could play an essential role in managing mental health in the future.

Researchers have also transferred microbes via 'faecal microbiota transplants' from one organism to another. These studies show a link between microbes and behavioural traits such as anxiety-like behaviours and anhedonia – or the inability to feel pleasure.[10] In essence, specific behaviours have been transplanted from one organism to another via microbes. So, taking host–microbiome interactions into account could provide a far richer and more accurate explanation of behaviours like these.

Are you the kind of person who forgets important dates? Or perhaps the type of person who forgets to express affection? Well, your resident microbes could potentially affect elements of your cognition, such as memory. Microbes can affect the host's recall of food locations in animals. In wild animals, this could have

Could the microbes from individuals without anxiety and depression be transferred to anxious and depressed individuals to reduce their symptoms?

important implications for health by affecting the creature's dietary intake. In humans, if something affects memory, it could theoretically cause relationship issues – for example, if you regularly forget an important date. Now, I'm not saying you can blame this on your microbes, but I'm not saying you can't either!

Understanding the potential role of host–microbe interactions in our behaviour and cognitive biases could help us understand people's complex behaviours and open up new realms of empathy. Moreover, if part of our perception and intuition (i.e. System 1 thinking) is influenced by our microbes and environment, could this change how we view those very concepts? Or even the way we view one another?

In the future, could we see a change in our levels of empathy for decisions that may be 'out of our control', or developments that help regulate impulses that would ordinarily lead to negative actions?

This reminds me of another Robert Sapolsky book, *Behave: The Biology of Humans at Our Best and Worst*.[11] Sapolsky says, 'any given type of explanation [for a behaviour] is the end product of the influences that preceded it'. He refers to the chain of biological, environmental and social events that lead to a given behaviour. For example, what happens in the person's brain a second before a behaviour occurs? Then what happened a second before that to cause the central nervous system to trigger the behaviour? Perhaps a startling sight or sound, or a potent smell. And what hormones circulating hours or days before affected our response to the stimuli? We can keep zooming further out, and we end up with a long and convoluted chain of events that ultimately lead to a given behaviour.

We can consider the holobiont blindspot, as with Sapolsky's biological chain of events, when thinking about the psychological concept of 'free will'. This, as Sapolsky regularly postulates, has implications for the notions of responsibility and punishment. Indeed, changes to certain parts of our central nervous system, such as the prefrontal cortex in the brain, can, as Sapolsky suggests, 'produce an individual capable of differentiating right from wrong but who, nonetheless, is organically incapable of appropriately regulating their behaviour'.[12]

Could our microbiomes affect our perceptions, actions and intuition by regulating our impulses? Should we consider this when debating the notions of free will and determinism? Perhaps with additional research, we'll find the answers and change our behaviour accordingly… unless our microbes have other ideas.

CHAPTER 8

The Glue that Holds Our Ecosystems Together

'If you don't like bacteria, you're on the wrong planet.'
—Stewart Brand

Microbes are the glue that holds our ecosystems together. To contemplate this, we only need to think about the many ecological functions that microbes are involved in: plant and animal health, communication between plants and between other kingdoms of life, nutrient and water absorption, hormone production in both plants and animals, pathogen resistance, climate regulation and pollination. The list goes on and on.

Let's start from the ground up. Microbes help lay the foundations for plant communities to survive and thrive. Bacteria in the soil can be single free-living cells, small groups or large populations living in gooey biofilms. These biofilms are formed when bacterial cells attach to and consume organic matter. When they digest it, they produce sticky and powerful glues. These glues attach to the surrounding soil particles, and over time, biofilms are formed. This sticky situation helps in the formation of soil and prevents soil erosion. The bacteria living in these biofilms help to form aggregates in the soil. They are vital to a wide range of soil functions, from securing nutrients and preventing erosion to reducing water stress and supporting plant root systems. They provide tiny houses for the invisible biodiversity. So, microbes are literally glueing the ecosystem together from the ground up.

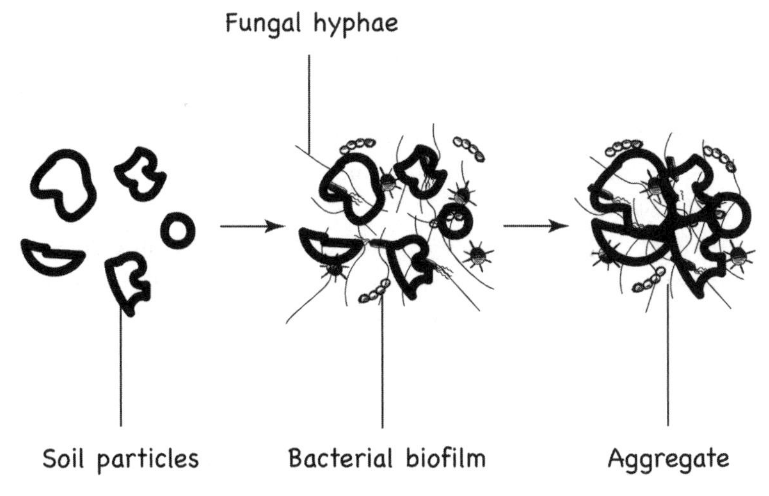

Microbes help glue the soil together, preventing erosion.

Fungi also have roles in building clumps in soil. They, however, act more like nets than glue. Fungal mycelial networks physically attach to smaller clumps of soil and wrap them up to form larger clumps. Even when soil fungal hyphae (the branching filaments that make up mycelium) begin to decay, they support the soil by secreting sticky substances from their cell walls. These substances can remain in the soil for half a century.

Microbes also play critical roles in cycling nutrients, making them available to plants. They supply nutrients to plants by, for example, converting nitrogen, sulphur, phosphorus and other compounds into plant-friendly forms. Microbes do this by degrading organic matter: basically anything that contains carbon, such as faeces or the remains of animals and plants. You can think of the microbes as the farmers, recycling plant operatives, miners and supermarkets of the plant world.

Imagine if soil microbes were wiped out immediately, gone in the blink of an eye – no more soil microbes on the planet. If microbes weren't here to break down the complex organic compounds, all the remains of plants and animals – including their waste – would rapidly build up. Very soon, we would all be wading knee-deep in the stuff. Plants are reliant upon microbes to survive. If microbes were wiped out, plants would no longer draw in vital nutrients and convert them into useful chemicals. They would rapidly lose all capacity to produce energy via photosynthesis, and would swiftly die. All other organisms that depend on plants to survive would soon be cursed with the same fate. If this doesn't highlight how incredible and underappreciated microbes are in our ecosystems, I don't know what will!

So, we know microbes support soil structure, provide vital nutrients to plants and prevent us from wading knee-deep in gunk. However, they are also involved in plant communication and are particularly active in the root zone or 'rhizosphere' – more on this in the following chapter on microbes and trees. But let's jump out of the soil now and look at the above-ground portion of plants, known by the scientific community as the 'phyllosphere' – from the Greek derivative *phyllo* for 'leaves', and *sphere* for 'globe', it's the 'leaf world'.

The phyllosphere microbiome has parallels with the human microbiome. For instance, the microbial community here plays a role in immunity and resistance to pathogens. Friendly microbes living on the plant leaves can help stave off infections by (1) inducing a plant immune response, (2) outcompeting incoming opportunistic pathogens, and (3) producing antibiotic compounds. Like in the human body, many phyllosphere microbes act like tiny chemical factories, producing stimulants that induce the expression of pathogen-resistance genes in the plant – this can result in a local or whole-plant response. Some bacteria can degrade the cell walls of pathogenic fungi. This has the effect of reducing fungal infections. The phyllosphere microbiome also plays a role in outcompeting pathogens. For example, *Sphingomonas* strains can inhibit the plant pathogen *Pseudomonas syringae* by vigorously competing for the same nutrient resources.[1]

Imagine one medieval army (the pathogens) storming the castle of another army (the protectors). They both want the same thing – the riches and rights of the kingdom – but the protector army is strong and is unwilling to give it up, despite the forthcoming assault. A phyllosphere microbiome with a greater number of species is linked to improvements in the plant's capacity to suppress diseases. An increase in species diversity often corresponds to more ecological functions being fulfilled. Think about our medieval armies again. If our protector army has a greater variety of soldiers and weapons, they may have a better chance of defending the castle. In contrast, the castle may fall to the enemy if they only have foot soldiers and not archers, crossbowmen, spearmen and gunners.

Finally, phyllosphere microbes can produce antibiotics. It is thought that bacterial antibiotics on plant leaves can be broad-spectrum, acting against pathogenic bacteria and fungi. Similar activities occur *inside* the plant too. This is known as the 'endosphere' (the 'inside world'). The microbes in the endosphere have a variety of roles to play in the plant's immunity. Microbes also have roles in the plant's stress tolerance. For instance, studies suggest biofilm formation on plant leaves could protect the plant from desiccation and ultraviolet radiation. In addition, some plants are more tolerant of heat stress when inoculated with fungi that live inside the plant. And plant microbes can remediate chemical pollutants, increasing the plant's tolerance to toxins.

Pollination

Recent studies suggest that pollination activity – for example, when bees, flies and beetles hop from one flower to the next – is partially influenced by the microbes residing in the flowers. Chemical cues produced by flower-dwelling microbes influence a bumblebee's feeding decisions.[2] Microbes consume the abundance of sugars in the flower's nectar and produce a rich bounty of chemicals. Clouds of these volatile compounds sail through the air, and bumblebees can detect them and make informed decisions about where to feed!

Dr Robert Schaeffer and his team at Utah State University discovered this back in 2019. First, they produced contamination-free artificial nectar and inoculated it with either a bacterium or fungus commonly found in flowers. The researchers then placed the nectar solutions in separate arms of a Y-shaped choice chamber. This allowed the bumblebees to choose which arm of the chamber they preferred. The bees spent considerably more time in the arm with the bacterium, suggesting they preferred the volatile compounds given off by it. When they placed the bees in individual chambers and gave them just one choice, they consumed more fungi-spiked nectar. The researchers think this is because the bacterium produced twice as much of an attractive compound, thus outcompeting the fungi compound when nearby. Imagine walking down the high street to get some lunch, but you're not sure where to go. There are two bakeries nearby but only one has that wonderfully comforting baked bread smell wafting out of the doorway and into the street. I don't know about you, but that's the one I would make a beeline for (aha, pun not intended). It seems the bees work in a similar way.

So, microbes influence how attractive a plant is to its animal pollinators. They do this by altering the plant's visual, gustatory or olfactory cues. Some bacteria called *Acinetobacter* in flowers can send over a chemical signal to the flower's pollen and hijack its system, as discussed earlier.[3] The bacteria effectively tell the pollen to open its doors from the inside and release nutrient-rich compounds for the bacteria to consume. We can often find *Acinetobacter* on pollinators such as bees and flies. Therefore, do pollinators also benefit from this microbe-induced germination?

Climate regulation

Microbes also have roles in climate regulation. Marine microscopic phytoplankton (near-invisible, free-floating plants) are responsible for 50% of the global photosynthetic CO_2 fixation and oxygen production. This is despite phytoplankton only comprising around 1% of global biomass.[4] Marine phytoplankton are distributed over a larger surface area than terrestrial plants. Their lifecycle is often days rather than the decades of, for example, trees – arguably more complex organisms. Because of this fast turnover, phytoplankton can respond more rapidly to climate variations on a global scale. Rising atmospheric CO_2 can increase phytoplankton productivity and thus CO_2 fixation, but only if nutrients are unlimited. This is an issue because a rise in temperature, for example due to human-made climate change, can reduce the transportation of nutrients from the deep water to the surface where phytoplankton reside. This results in a loss of phytoplankton productivity, thereby further affecting the climate. There are conflicting studies that show global phytoplankton productivity both increasing and decreasing, which makes this realm challenging to talk about. Still, it is thought that climate change could soon diminish phytoplankton biomass, and thereby reduce global CO_2 uptake.[5] This could potentially exacerbate climate change.

Many other microbes, including bacteria, archaea and viruses, play critical roles in nutrient cycling and climate regulation in the deep ocean abyss. They influence gas exchange and the ocean's carbon sequestration capacity. A single litre of seawater can contain 10 billion bacteria and 100 billion viruses, and the density is far higher in the top 15% of the deep ocean sediment – which contains around 13% of all the bacteria on the planet. It's hard to convey just how dense and buzzing with energy the seabed really is. It's like a vast and continuous food festival on a microscopic scale; tiny bodies bumping into each other, buying and selling products, exchanging energy – some making a mess, others cleaning it up. Swathes of cells talking to each other, making informed 'decisions'. The impact of microbes on the rest of the ocean food web is immense.

Let's now think about the role of terrestrial microbes in climate regulation by jumping back into the soil. Soil microbes regulate the organic carbon stored in soil and the amount released back into the atmosphere. They indirectly influence plants' carbon storage by providing nutrients that regulate productivity. On the other hand, plants remove CO_2 from the atmosphere through photosynthesis and create organic matter that fuels our ecosystems in a harmonious relationship with microbes. These opposing processes are primarily responsible for climate regulation. Unfortunately, climate change is expected to accelerate the release of CO_2 into the atmosphere by promoting higher rates of organic matter decomposition.[6]

Rain-makers

The common saying 'it's raining cats and dogs' should probably be replaced with 'it's raining microbes'. That's because the sky is teeming with a group of bacteria that can actually make it rain – a process known as 'bioprecipitation'.[7] Initially, these bacteria form colonies on plant leaves. Then winds sweep them into the sky, where tiny ice crystals cling to them. Unlike other particles in the atmosphere, the ice-forming bacteria have evolved to increase the temperature at which ice crystals form. As the water molecules build up around the bacteria in the atmosphere, eventually it starts raining, and the microbes fall back down to the land in the rain droplets. The rain-making bacteria likely benefit from this process because the rain droplets disperse their cells into new habitats, much like plants rely on windblown pollen. The most common of these rain-making bacteria is the plant pathogen *Pseudomonas syringae*. Incidentally, many ski resorts use freeze-dried proteins from rain-making bacteria to stimulate ice formation in snowguns. So, you may have *P. syringae* to thank for the quality of the powder on your downhill runs!

Some of the rain-making bacteria can cause damage to plants, but on the other hand, they can also be part of a constant feedback loop between terrestrial ecosystems and clouds. Reducing the amount of rain-making bacteria on plants could affect climate regulation. For example, overgrazing or habitat clearance could potentially reduce these bacteria, thereby decreasing rainfall. This has implications for the health of the ecosystem in subsequent years. So, the next time it's raining, perhaps take a moment to wonder whether these remarkable bacteria made it happen.

Sticky situations

Microbes can be literal adhesives in some cases. The world's strongest known glue is made by a bacterium called *Caulobacter crescentus*. This water-dwelling bacterium secretes a sugary substance so powerful and sticky that just a tiny amount can withstand the 'pull from lifting multiple cars at once'. It may have an adhesive force of over 2 tonnes per square centimetre.[8] Others, meanwhile, say it takes around 70 newtons per square millimetre to pull a single bacterium away from a surface. It is remarkably strong, in any case.

The bacterium sticks to surfaces to endure the harsh conditions of the oceans and remain stationary for long enough to collect

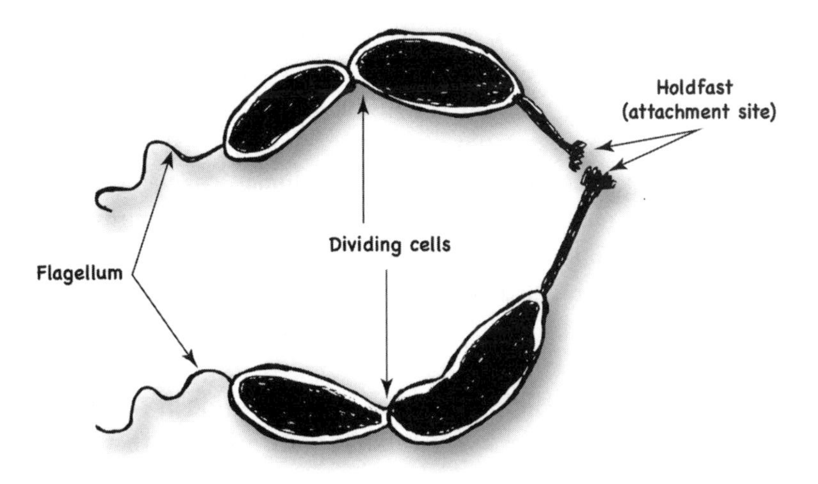

The incredibly sticky *Caulobacter crescentus*.

passing food items. Once *C. crescentus* adheres to a surface, it forms a biofilm. As discussed previously, within biofilms, many other bacteria organise themselves into complex, tight-knit communities. When conditions are favourable, the cells produce adhesives just like *C. crescentus* does to strengthen their attachment to surfaces.

This communal lifestyle occurs throughout our ecosystems and benefits the microbes by protecting them from predation and other stressors. Biofilms support a wealth of microbial diversity and fulfil essential functions in many ecosystems. We now think biofilms are the dominant mode of microbial life on Earth.[9] Biofilms are fundamental to coastal marine ecosystems. They provide a rich food source for benthic grazers and a 'settlement' substrate for corals, which are key ecosystem engineers. Biofilms help purify water and store carbon, and they are vital players in nutrient cycling in freshwater ecosystems. Biofilms are also important in agriculture, where they help stimulate plant growth and control soil borne pathogens.

You may notice that biofilms receive a disproportionate amount of bad press. This is because they can also be detrimental to humans. For example, they can form impenetrable communities inside the human body in certain conditions. When this happens, some biofilms can cause diseases that are difficult to treat because of the protective matrix that surrounds the community. Nonetheless, the resilient nature of biofilms will likely play a pivotal role in our adaptation to stressful phenomena like climate change and soil erosion. We already know that bacterial and fungal biofilms help maintain the equilibrium in soil, providing nutrients to plant roots and supporting soil structure. Researchers are now investigating how to inoculate soils with biofilms to increase the resilience of the ecosystem.

Sticky biofilms will likely play an important role in ecological restoration in the future; so, microbes really are the glue that holds our ecosystems together.

CHAPTER 9

Microbes and Trees

'Think of the trees and how simply they let go, let fall the riches of a season, how without grief (it seems) they can let go ... Learn to lose in order to recover, and remember that nothing stays the same for long ... Let it go.'
—May Sarton

It was a crisp autumn day. The leaves were flittering in the breeze. The squirrels were scurrying around to cache the last nuts ready for the long winter ahead. I looked down at the carpet of the beautiful but decaying foliage beneath my feet – a marvellous array of colours: bronzes, browns, scarlets and yellows. Out of the corner of my eye, I noticed a couple of leaves with small green patches as though the process of senescence was incomplete. Generally, when autumn arrives with shorter daylight hours and colder temperatures, deciduous trees prepare for winter and draw nutrients and water away from the leaves and back into the stem and roots. This reallocation of resources is what defines the term 'leaf senescence' – or the ageing of leaf cells. *Abscission* is the term used for when leaves fall from the branches. The word comes from the Latin for 'cut off'. Incidentally, the process of abscission in leaves involves the competing actions of two plant hormones – ethylene and auxin. Auxin inhibits leaf abscission, whilst ethylene promotes abscission. When autumn begins and the characteristics of the changing season trigger leaf senescence, the secretion of auxin reduces, and the base of the leaf then becomes sensitive to the ethylene hormone. During the final 'shedding phase', ethylene causes the cell walls at the base of the leaf to degrade. This is why leaves fall to the ground, creating that wondrously comforting and colourful carpet, preceding the winter sludge.

The small green patches I noticed on some of the bronze leaves indicated that an island of chlorophyll (the green pigment) was holding on despite the tree's prior efforts to draw nutrients back into the stem, which involves the degradation of chlorophyll in the leaves. I then remembered these patches were known as 'green islands'.

The green islands are seemingly caused by insects called leafminers – which are the larvae of various species of moths. They try desperately to prevent leaf senescence, thus prolonging their pantry of food on what would otherwise be a dead, decaying leaf. Although this green island trait was known for several decades, the causal mechanism remained elusive. In the late 1960s, entomologist Lisabeth Engelbrecht set up an experiment to study leafminers on birch and aspen.[1] She aimed to test her hypothesis that leafminers caused the green islands by either secreting special saliva or cutting the leaf veins that would deliver the chemicals to trigger senescence. Engelbrecht discovered that green islands contained high amounts of a particular plant growth-promoting hormone called cytokinin. Engelbrecht and colleagues found that the saliva in the labial glands (around the mouth) of the leafminers also contained large quantities of cytokinin. This was quite an exciting finding. However, it was strangely followed by a long hiatus in leafminer and cytokinin research until the mid-2000s. As we know from many ecology studies, when one member of a given kingdom (e.g. an insect) interacts with a member of another kingdom (e.g. a plant), the underlying mechanism often involves microbes. The green island story is no exception.

Scientists eventually demonstrated that *Wolbachia* – a ubiquitous group of bacteria occurring in 60% of the planet's insect species – were helping the leafminers in their green island adventure.[2] French biologist Dr Wilfried Kaiser and his team later treated leafminer larvae with antibiotics to remove the *Wolbachia* bacteria. Incredibly, the researchers discovered that the antibiotic-treated larvae could no longer produce green islands and the cytokinin levels reduced significantly. This is another example

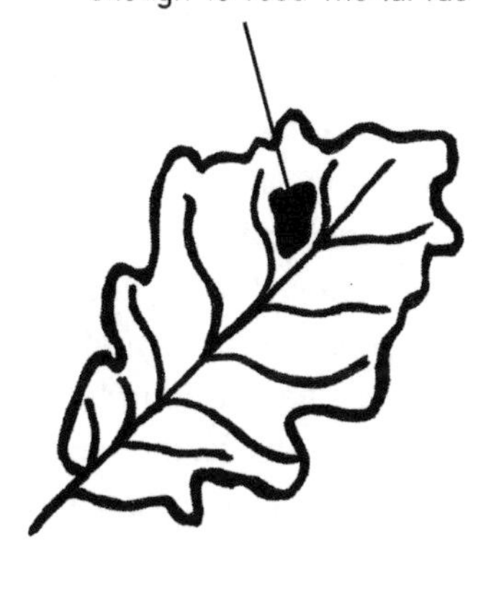

Green island and leafminer trail.

of a co-evolutionary relationship whereby the leafminer and the *Wolbachia* both benefit from a prolonged food source.

These relationships between microbes, trees and insects are wide-ranging in nature and likely play an essential role in maintaining the integrity of the ecosystem. Microbes also have intimate and specific relationships with trees that ultimately ensure the tree's survival.

Professor Suzanne Simard grew up in the forests of British Columbia in Canada.[3] She would lie on the forest floor and stare up at the giant tree crowns as a child. Her grandfather was a horse logger who selectively cut cedar poles from the inland rainforest. Suzanne describes how her dog Jigs once fell into the outhouse pit (also known as an outdoor toilet!) and was trapped. Her grandfather ran up to the pit with a shovel to rescue the poor dog. As he dug the forest floor, Suzanne became fascinated with the dense web of roots, mycelium, soil and minerals revealed by the spade. She realised this combination of elements was the foundation of the forest.

She wanted to know more and studied forestry, but soon became conflicted by her role in an industry that clear-felled and sprayed chemicals on trees, and replaced them with commercially valuable conifers. At the time, scientists had recently discovered that in lab conditions, one pine seedling root could transfer carbon isotopes to another pine seedling root. But this was in the lab. Suzanne began wondering, could this happen in a natural setting? She eventually conducted experiments in the Canadian forests, where she encountered grizzly bears that chased her off and endured the wet and isolated temperate rainforest conditions. She borrowed a bounty of high-tech scientific gear from the university, including a Geiger counter. She bought some 'really dangerous stuff' – syringes packed full of radioactive carbon 14 CO_2 gas and some bottles of stable carbon 13 CO_2 gas. Suzanne then placed bags around birch and Douglas fir saplings and injected the radioactive isotopes into the birch bags. She hoped the birch trees would take up the gas and shuttle it to their neighbouring species, the Douglas firs. She joked that the grizzly bear showed up again, and she jumped out of her skin, accidentally waving the radioactive syringes around whilst being eaten alive by mosquitoes. She dived into her truck, and at that very moment, she thought, *This is why people do lab studies.*

After an hour, the grizzly had left, so she removed the bags from the birches and ran her Geiger counter over the leaves. The moment of truth. *Kkhhhhh!* Excellent: that sharp, fuzzy radioactive noise never sounded so sweet. The birch had taken up the C-14 radioactive isotopes. She went over to the fir tree and ran her Geiger counter over the leaves. *Kkhhhhh!* Wonderful! The birch had indeed shuttled the isotope over to the fir tree. They were talking! The fir asked for carbon, and the birch would say, 'Sure thing, here you go!' The Douglas firs sent the other isotope (C-13) to the birch trees. A two-way conversation! Amazingly, when the firs were shaded and could no longer take as much carbon dioxide from the atmosphere, the birches would send even more carbon over to the firs. When the birches were leafless in the winter, the firs would return the favour and send more carbon over to the birches. The two species were interdependent.

Suzanne had discovered evidence of strong cooperative bonds between trees in the forest, whereas the prevailing historical view was that of competition – different species competing for light, water and nutrients. This cooperation was not just cooperation between the trees. It was cooperation between the trees and an underground world of infinite biological pathways connecting the trees, allowing the forest to behave like a single organism. It turns out that the trees were not just conversing using the language of carbon, but also with other gases, hormones, vibrations and water – in other words, they were using a vast reservoir of information to communicate with and support each other.

Scientists had previously thought near-invisible mycorrhizal fungi mediated the transfer of various forms of information between trees. The fungi exist as tiny threads called *hyphae*. These hyphae interconnect into a complex network called *mycelia*. This mycelial web can be so dense that there can be hundreds of kilometres of mycelium under a single footstep, in a healthy forest at least. Mycorrhizal networks connect different individuals but also different species – like the birch and fir. This dense network teeming with bi-directional information flow has been likened to the internet, and the colloquial term 'wood wide web' is widely used to reflect this phenomenon. Mycorrhizal fungi connect the plants in a few ways. One Is through ectomycorrhizae ('outside fungi') which surround the outside of the tree roots. Another is through endomycorrhizae ('inside fungi'), which grow inside the tree roots where the hyphae wedge themselves in between the cell wall and cell membrane – kind of like squeezing themselves in between a bike tyre and an inner tube. Mutualistic fungi called arbuscular mycorrhizae also live in tree roots and extend their hyphae into the wood wide web to facilitate nutrient uptake from the soil.

Suzanne says there are vital 'hub trees' in the network. These are more fondly known as 'mother trees' because it turns out they nurture their young – the smaller trees growing in the forest understorey. Through the dense microbial network, the hub trees can connect to hundreds of other trees in the forest. The hub trees will shuttle excess carbon to the seedlings, which increases the seedlings' chances of survival. The trees can also warn each other about stressors and enhance each other's resistance to pathogens via this complex network.

In addition to mycorrhizal fungi, another group of mutualistic microbes involved in tree health and communication are the *mycorrhiza helper bacteria*. These diverse bacterial groups can enhance mycorrhizal function, growth, nutrient uptake, soil conductance and pathogen resistance. In one study, the bacteria helped fungi to form a mutualistic relationship with the tree. Scientists hypothesise that the 'helper bacteria' help trees and other plants defend themselves against pathogens by enhancing nutrient uptake and allowing plants to allocate more resources to defence mechanisms. So not only do vast networks of microscopic soil fungi help trees communicate, grow and defend themselves, but microbes from an entirely different kingdom – the bacteria – help these fungi form mutualistic relationships with the trees. A spectacular evolutionary arrangement formed over thousands of years of cooperation between the micro and the macro worlds.

We can learn a considerable amount from these ancient and complex relationships, for they hold a profusion of wisdom. But above all, we must endeavour to protect them for the future generations of trees and forests, along with the myriad species they support.

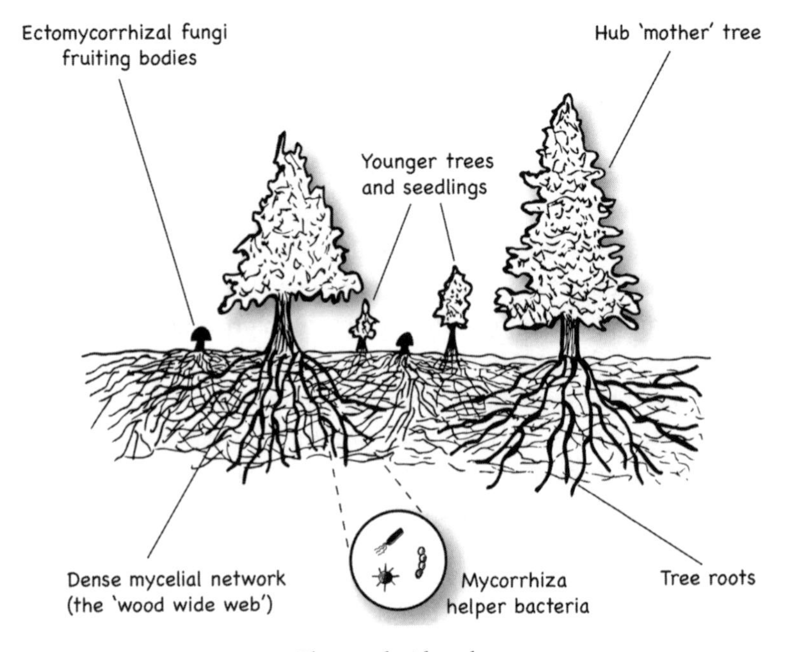

The wood wide web.

I find myself arriving back at Dawkins's theory of the *extended phenotype*, whereby the genes that influence a host's behaviour tend to be maximised because of the behaviour itself. Recent evidence suggests that fungal communities are responsible for differences in drought tolerance between host trees. The microbial denizens in the soil near the tree roots are partially under the plant's control, and vice versa. Therefore, we can view the make-up of the microbial communities as an extended phenotype of the host tree. The primary benefit is that the fungi help the trees adapt to the increased drought stress caused by a changing climate.

A tree's ability to adapt is also enhanced by associating with microbes from the local environment. Plants grown in close association with microbes from stressful environments can outperform plants grown with microbes from less stressful environments when transplanted to the stressful surroundings. This means the microbes from the stressful environments are adapted to the severe conditions. But not only this; microbes can also pass this adaptive advantage on to the plants – another symbiotic relationship between different kingdoms of life. Imagine living in a temperate country like the UK, and then going on a trip to the Saharan desert with a friend from home. You probably wouldn't survive for long on your own. However, now imagine you went to the desert with a local guide, who was resilient and had all the relevant knowledge needed to survive the harsh environment. You would have a much better chance of surviving because you were with this local guide, who passed their knowledge on to you.

In science, we often limit investigations to single organisms, compounds or atoms. However, it is vital that we also take a holistic approach to connect the dots, see the relationships, understand the complexity and (pardon the pun) 'see the woods for the trees'. Many trees, such as willows, tolerate pollution and contamination in the soil, including hydrocarbons and heavy metals. Cleaning soil with the help of trees and plants is known as 'phytoremediation'. However, it is not just the willows working alone but an intricate mutualistic symbiosis involving invisible friends. Evidence suggests the trees tolerate the contamination by shuttling sugars to symbiotic fungi surrounding the roots. In turn, the fungi provide nutrients to pollution-gobbling bacteria.[4] It's like when a company gets the credit for an invention or movement. But really, the company provides a resource to their managers – in the form of a hefty pay packet – who, in turn, hire an external 'think tank' to help generate the innovative ideas. They all have vital roles individually, but the magical ending arises through a well-orchestrated collective process.

An important tree called the common alder (*Alnus glutinosa*) grows in Europe. They can grow to a height of 28 metres and typically live for around 60 years. The bark of the alder is dark and fissured. It's often covered in lichens and mosses, and it supports many insects – beetles, butterflies and moths – including the charming alder kitten moth. The alder is monoecious, meaning you'll find both male and female flowers on the same tree, as opposed to dioecious, where individuals have either a male or a female reproductive organ. Its natural habitat is moist ground near rivers and lakes, along with cool, wet marshy grasslands. Like birch, alder is a pioneer species because it can seed and quickly colonise bare ground. It also has a magic trick up its sleeve. It can rapidly colonise nutrient-poor soils where few other trees would thrive. This provides alder with a survival advantage.

This ability only occurs because of the tree's association with *Frankia alni*, a symbiotic nitrogen-fixing bacterium. *F. alni* forms a symbiotic relationship exclusively with trees in the genus *Alnus*. The *F. alni* bacteria 'infect' the root hairs of the alder and begin forming nodules around the roots. They produce enzymes that help fix nitrogen from the atmosphere, thus providing valuable nutrients to the tree. Studies suggest *F. alni* provides 70–100% of the tree's nitrogen requirements.[5] Because of their mutually beneficial relationship with *F. alni*, alder also improve the fertility of the soil in which they grow. Evidence suggests that the Incas in pre-Columbian South America may have exploited other alder tree species.[6] They probably used the tree and its symbionts to increase soil fertility and stabilise terrace soils in their upland farming systems. Additionally, this process enables other successional species to become established – another reason alder is considered a pioneering species (but really, it's the alder and its invisible friends).

While researching the topics for this chapter, I found myself chatting to a good friend, Harry Watkins, director of St Andrews Botanic Gardens in Scotland.[7] Harry and I have worked on several microbial ecology projects together; we both did our PhDs in the same department at the University of Sheffield. Harry told me about a tree at the botanic gardens that was heaving with lichens. He showed me a photo of the tree, and it looked like it was wearing a dense and fluffy jade-coloured coat – which in fact was *Cladonia*, also known as reindeer lichen. The tree is a *Sorbus sargentiana*, a Sargent's rowan, native to south-western Sichuan and northern Yunnan in China. Or at least that's what the black label and database imply. Harry says you can see the tree has been grafted about 1.5 m high, and whilst the top half may be Sargent's rowan, the bottom half is clearly something else – most likely a *Sorbus aucuparia*, our native rowan tree or 'mountain ash'. Tree grafting is a horticultural technique in which plant tissues are fused to continue their growth as one united organism. It is usually performed on trees of the same species; however, occasionally, grafting one species onto another species of the same genus is carried out. Indeed, this is the case with Harry's *Sorbus* tree(s). Imagine grafting a blackbird's head onto the body of a song thrush. Yeesh. The whole *Sorbus* organism was recorded as a single taxon, when very clearly it's a mixture of at least two species of rowan plus all its microbial and epiphytic symbionts – the organisms such as lichens and mosses that grow on the plant. This conversation led us to question the blurry lines between one species and another.

What is a species anyway? At first, it may seem relatively straightforward. However, as naturalist Charles Darwin said in *On the Origin of Species*, 'Nor shall I here discuss the various definitions which have been given of the term species. Not one definition has as yet satisfied all naturalists'.[8] Later, in a letter to Joseph Hooker, he said, 'It all comes, I believe, from trying to define the undefinable'.[9]

The human brain is acutely attuned to recognising patterns. Thus, it is also prone to the cognitive blindspot of finding meaning where it does not exist. It seems we cannot resist the temptation to classify things, a phenomenon deeply rooted in cognition but also in language – perhaps in some cultures more so than others. Whilst many concepts of what a species is have been proposed – from the *typological species concept*[10] whereby similar-looking organisms are grouped into a species, to Ernst Mayr's *biological species concept*[11] whereby a species is a group of organisms that can reproduce with each other, but not with other groups – it isn't easy to imagine that an entirely satisfactory definition will ever be found.

Defining a species becomes even more complex when we account for the trillions of microscopic residents living in and on an organism in a close-knit relationship; the microbes play critical roles in the organism's health and survival, potentially even their sexual preferences and thus their reproductive activity. However, one thing is for sure: Harry's *Sorbus* is not *just* a *Sorbus*. Thinking back to how the tree was grafted from one species and onto another, also known as 'xenografting', we wonder if this had something to do with establishing the tree's ultra-dense coat of *Cladonia* lichen. Perhaps the answers lie in studying the tree's microbiome… or, indeed, its graftobiome?

Not all microbes are friends, and some cause diseases. Understandably, we often view pathogens with hostility. Yet they are part of a normal functioning ecosystem and, like any other organism within that system, their goal is to live and procreate. They can wreak havoc, though, especially if helped along by the activities of humans, as highlighted by the ongoing human-associated COVID-19 pandemic. Indeed, the ongoing degradation of ecosystems means we are living in good times for 'bad' microbes, and bad times for 'good' microbes.

Dutch elm disease is caused by the roll-off-the-tongue fungus *Ophiostoma novo-ulmi*. We refer to this as 'vascular wilt' because the fungus blocks the tree's vascular system, which is responsible for transporting water and nutrients. This causes the branches to wilt and die. Elm bark beetles spread the fungus. We accidentally imported them into the UK from Canada in the 1960s. By the 1970s, over 90% of elms – or 25 million trees – were lost.[12]

Another tree disease currently wreaking havoc in the UK is ash dieback. Once again, a fungus, namely *Hymenoscyphus fraxineus* causes this disease. However, unlike the Dutch elm fungus, ash dieback fungus spores can travel in the wind. The fungus was imported to Europe from Asia on commercial ash saplings. It then spread like wildfire on the breeze.[13]

It is thought that some trees can resist these few bad apples with the help of… you guessed it, invisible friends! Microbes play integral roles in the health of trees, much as they do in humans. From this holistic perspective, the bacteria, viruses, archaea, fungi, algae and protozoans that form the tree holobiont are vital to its ability to adapt to incoming diseases. More specifically, scientists recently discovered signs that certain bacteria in the leaves of ash trees might provide some protection against the ash dieback pathogen. The researchers studied the microbes in leaves from susceptible and infection-free ash trees. They then isolated groups of bacteria in the lab and found some bacteria suppressed the growth of ash dieback fungus. The researchers said these bacteria might prevent the penetration and spread of the fungus.[14] A follow-up study is planned to assess whether the bacteria act by directly antagonising the fungus, by outcompeting it, or by inducing resistance in the tree. Fingers crossed for favourable results.

Another study found that horse chestnut trees with a disease called 'bleeding canker' harboured a lower diversity of microbes in the bark. The researchers are unsure whether the disease (caused by the bacterium *Pseudomonas syringae* – the 'rain-maker') induced this reduction in microbial diversity, or whether trees with a more diverse microbiome have greater resistance to the disease.[15] Determining this will be necessary if preventative treatments, for example 'tree probiotics', are to be developed.

Human activities and the by-products of urbanisation are a threat to the health of trees. For example, air pollution can weaken trees and make them more susceptible to pathogens. Pollution can also alter the microbiome of tree pollen, potentially increasing its allergenicity – that is, humans may become more vulnerable to pollen allergies because of pollution's impact on the pollen microbiome.[16] Once again, this points to the need for a transformational shift towards holistic approaches to living. Our behaviours and actions can have dramatic impacts on our surrounding environments.

We're in this together – from the microbes to the trees, to the birds and the bees. Let's move the dominant parasitic relationship with nature towards a mutualistic one. Both are symbiotic, but the latter is the only way to ensure all life in its current manifestations flourishes well into the future.

CHAPTER 10

Rewild, Regenerate, Restore

'Walking. I am listening to a deeper way.
Suddenly all my ancestors are behind me. Be still,
they say. Watch and listen. You are the result of
the love of thousands.'
—Linda Hogan

The *Dirty Thirties*. The *Dust Bowl*. These are two names given to a period in the 1930s in the United States Southern Plains region. This was to do with the effects of soil and agricultural practices, along with a severe and prolonged drought at the time. After enormous bloodshed, during which over 600,000 soldiers died, the United States Civil War or the 'war between the states' ended in 1865. During this era, a series of federal land policies and acts enticed pioneers onto the Great Plains. The Homestead Act of 1862 provided settlers with 160 acres of public land to farm, and subsequent acts in the early 1900s coaxed an influx of inexperienced farmers to settle on the plains.[1]

Many of these settlers, along with politicians and scientists of the time, truly believed the 'rain follows the plough' climatology theory proposed by land speculator Charles Dana Wilber:

> God speed the plow … By this wonderful provision, which is only man's mastery over nature, the clouds are dispensing copious rains … [the plow] is the instrument which separates civilization from savagery; and converts a

> desert into a farm or garden. … To be more concise, Rain follows the plow.[2]

The basic premise of this theory was that human homesteading and intensive agriculture permanently altered the local climate, making the region more humid, fertile and lush as the population increased. Wow.

Hindsight is a wonderful thing, but it's safe to say this was an overwhelming catastrophe. The theory had a profound impact on the land in the United States, but it was also used to justify the expansion of intensive agriculture in Australia. Unfortunately, a series of wet years during this period fused this theory into the public's psyche. It has since been refuted and is now considered to be mere superstition.

Wheat prices plummeted during the Great Depression, and farmers ploughed vast swathes of land to harvest more as they tried to break even. But in 1930, a severe and prolonged drought hit the country. Crops began to fail, soils to erode. By 1934, roughly 35 million acres of settled and over-cultivated land had been rendered useless for agriculture.[3] Many millions of acres on top of this were rapidly losing vital topsoil. Powerful storms blew the lifeless and eroded soil around, hence the term Dust Bowl. The storms were so powerful, and the earth so eroded, that the soil was swept away from the Great Plains in the heart of the country, all the way to the Atlantic Ocean where ships were coated in the dust. These storms were known as 'black blizzards'.

Lights out. Great swirling clouds of soil dust would cover the sky, darkening the days. Often residents would spend hours shovelling the piles of drifted dust like snow in the winter. Dust would coat homes, inside and out. People developed respiratory issues, 'dust pneumonia' and difficulty breathing. Thousands likely died from the dead soil of the black blizzards.

'Soil' is the crucial word in this story. In 1937, United States president Franklin D. Roosevelt wrote, 'the nation that destroys its soils destroys itself'.[4] This remains a fundamental truth: soil condition is a profound indicator of the planet's health. In Western societies, in particular, we fail to place sufficient value on and establish deep relationships with the ground beneath our feet. As Professor Duncan Cameron said, 'the use of the word "dirt" to denote inferiority is an example of this disrespect for our land. Yet societies succeed and fail as a direct consequence of the value they place on their soils'.[5]

Microbes and mesofauna (e.g. worms and tiny insects) are vital members of the soil community. And a 'community' is just how we should view it. It is far more than just 'dirt'. It is the epicentre of terrestrial life, a bewitching microbial cosmos with unrivalled diversity. But if we disrespect the soil through unsustainable farming practices, this enchanting microscopic universe suffers.

It doesn't just end with the death of the microbes and mesofauna. As we've discussed in previous chapters, these invisible friends and tiny creatures are essential to the health and survival of the entire ecosystem – the plants and animals, including our own species, *Homo sapiens*. If we continue with our current intensive farming practices, we may still be able to feed the projected nine billion people on the planet by 2050, for a short period. But our life-support system – the biodiversity and its provisions – will soon collapse. A new, or rather *ancient*, way is needed: *nature's way*. A deep ecological understanding of the land is required to create what many would consider an unorthodox food production model. But nature's way is the orthodox model – it's been around for aeons! We have just learnt to work against it, creating new, damaging conventions.

Learning from deep ecological principles and traditional knowledge, we can reverse the negative trends; reverse our impact on our life-support system; reverse our attitudes and values so they are more reciprocal with the land. One movement that embraces these principles is regenerative agriculture.

Regenerative agriculture employs rehabilitation and conservation approaches to farming systems. It focuses on restoring and conserving soil health, increasing biodiversity and enhancing the integrity of the ecosystem. This is not a specific practice in itself, but rather a set of principles that underscore a variety of sustainable farming methods. Considerable attention is given to the life in the soil. As we know, our microbial friends in the soil play a vital role in nutrient cycling. And when we destroy these friends, nutrient cycling collapses. This means farmers turn to the use of synthetic chemicals to provide these nutrients – which only exacerbates the issue. Plants also need microbes for their health. If the microbial denizens and their ancient relationships are missing, the plants become ill. When this happens, farmers turn to fungicides to kill the pathogens. But these also kill many of the remaining soil microbes, which again exacerbates the issue. More and more plant diseases develop, and we turn to pesticides to control opportunist insects. But these also kill the predatory insects that would, under normal flourishing conditions, keep the opportunists in check. The downward spiral goes on and on, and every corner of the ecosystem feels the impacts.[6]

The nutrient density in the foods we produce has declined enormously in the last few decades. A British study found that between 1930 and 1980, the average calcium content in 20 different vegetables had declined 19%, the average iron content by 22%, and the average potassium content by 14%.[7] Some reports suggest you would have to eat four carrots today to get the same amount of magnesium as you did from one carrot in 1940. Or eat 26 apples to get the same amount of iron as from a single apple back in 1940.[8] The adage 'an apple a day keeps the doctor away' may have been true a century ago; now I'm not so sure.

I thought that to write half a chapter on regenerative agriculture, I first needed to visit a regenerative agriculture farm. I have a broad knowledge of soil ecology, and I've spent copious hours working on traditional farms in the past, lambing and driving tractors. Yet this was a different kind of agriculture. It requires a shift away from orthodox farming practices and a profound change of lifestyle and philosophy. The best way for me to fully grasp and instil the context and nuances involved in regenerative agriculture was to talk directly to the farmers, walk around to get a feel for their land and learn about their practices – straight from the horse's mouth, so to speak. I scoured the internet to find the nearest regenerative agriculture farms and sent out several emails. The first person to get back to me was Chris from Fanfield Farm in East Sussex.[9]

Fanfield is known as an 'ecological farm', and their website states, 'we grow delicious chemical-free vegetables, whilst working hard to regenerate the soils and environment around us'. This sounded perfect.

Fanfield is a small farm, and they only produce local food and use the older hand-tool methods. This means they don't need giant tractors or hundreds of acres of land. In their own words, Fanfield Farm produces all of its crops regeneratively. Historical food systems have damaged soils and their microbial communities so that even farming 'sustainably' is no longer sustainable. Chris says, 'We have arrived at the point where putting back what we take out is not enough. We need to be putting more back than we take, and hence we need to farm regeneratively.'

Although they are fervent advocates of organic farming, Chris believes their farming standards exceed those practised on traditional organic farms. He says, 'We farm with completely organic procedures and practices. However, we are not certified organic. We are regenerative. Firstly, we don't use any pesticides or chemicals; we use all-natural matter, soil and compost.' Chris continues, 'The main reason why we're not certified organic is that we disagree that farmers doing the right thing should be penalised. To have the organic certification status costs a lot, and to spray vegetables with chemicals and pesticides costs nothing. We believe this is backwards.'

On the morning of 15 December, I headed over the rolling hills of the South Downs National Park towards Fanfield Farm. I passed vast swathes of traditional monoculture farms. I set off early to take a minor detour, as I'd realised that the Knepp Estate was not far away. Knepp is a 3,500-acre estate just south of Horsham, West Sussex.[10] Since 2001, the estate has been devoted to a pioneering rewilding project. The area was once intensively farmed and the soil biology was decimated, as in many other regions of the UK and the world. This rewilding project takes a largely hands-off approach to managing the landscape, although it still uses grazing animals. Knepp's driving principle is to establish a functioning ecosystem where nature is given as much freedom as possible, instead of locking down an ecosystem in its current state to protect a particular species. This is radically different to many conventional nature conservation projects.

Over the last 20 years, the Knepp soil and its complex life-forms have been regenerated. Previously, the cattle on the estate were treated with ivermectin for internal parasites, but these chemicals were killing more than just that – they were also destroying the dung beetles and probably many microbes in the cows and the soil. The Knepp Estate now recognises that without soil microbes and small but visible soil critters – along with the soil structure to retain them – water and nutrients leach away, and the soil becomes compacted and prone to erosion. They say that soil degradation is so severe in the UK that in 2014, the *Farmer's Weekly* magazine announced we have only a hundred harvests left.[11] Knepp still needs to carry out microbiome analysis on their soil, but surveys of earthworms revealed that a total of 18 different species now inhabit the land. Earthworms are vital indicators of soil health, and Knepp had plenty. They compared their soils to neighbouring farmland under traditional practices and found disproportionately high numbers of earthworms in all areas of the rewilding project. They've also analysed cow dung across the estate. They found 23 species of dung beetles in a single cowpat. Dung beetles fill a vital niche in the ecosystem, drawing down organic matter into the soil. This can be great for many other organisms, including microbes. Far across the land, they've also identified vast eruptions of fungal fruiting bodies – mushrooms everywhere! This demonstrates the spread of mycorrhizal networks that transport water, nutrients and chemical communication signals to the different plants across the land. Many of the mycorrhizae also form intricate symbiotic relationships with other organisms, including several beautiful orchids, which in turn provide food and refuge for invertebrates. Rare species such as nightingales and purple emperor butterflies now breed there and common species are skyrocketing.

As I approached the Knepp Estate, the first thing I noticed was a sign saying *Wild Animals Roaming*. This was music to my ears. I wandered around the footpaths, and quite soon, I saw huge piles of dung scattered across the marshy grassland. The dung was from cows, badgers, deer, foxes and rabbits. I thought to myself, *I must be surrounded by thousands of dung beetles too*. This was an excellent sign of the health of the landscape. Larger animals, smaller animals and microbes work together in a close-knit ecosystem. Even if they were not aware of it, they were helping to restore the degraded soils. Another thing I noticed as I walked around was the sheer number of holes. There were small and large holes in the ground where voles, badgers and rabbits had been building homes; there were holes in the large veteran oak trees where woodpeckers had been hammering away and where owls probably rested by day. This was another great sign. As the soils begin to regenerate, the vegetation returns and plant–microbe interactions help to restore the foundations of the ecosystem further; shortly after, the insects, reptiles and mammals return, and so do the passerine birds as their insect-and-berry food source booms. As the seeds and worms return, so do the voles and moles, and the raptors are not far behind them. Knepp's largely hands-off approach allows the animals, plants and microbes to restore energy flows, competition, predation and cooperation. These are vital facets of a flourishing ecosystem, and our invisible friends play a pivotal role.

It was a pleasure to see the early signs of a successful rewilding programme. Here's to hoping they can sustain the approach well into the future. This will likely depend on whether the adjacent habitats can be protected from huge developments. Indeed, the District Council is considering a major and controversial housing development site adjacent to Knepp as we speak. This could severely impact this jewel in the rewilding crown by destroying wildlife corridors, increasing pollution and affecting the groundwater. Conserving and restoring our natural environments is a constant challenge, but one that we must all rise to – there's no other option, there's 'no planet B', and the species we share this planet with deserve respect.

I continued my journey down the winding roads of Sussex to Fanfield Farm. I was greeted by Chris, who, along with his wife Emily, owns the farm on a long-term lease basis. He used to work in marketing and ditched his career a couple of years ago to pursue a more fulfilling and outdoor lifestyle. The farm only covers a few acres of land, but Chris and Emily use every square metre wisely, not just for crops but for wildlife too. Chris took me on a tour around the farm. They currently live on the land in a converted static caravan they extended, so it's more of a log cabin-caravan hybrid. The smell of slow-burning wood from the chimney was delightful. Chris took me past several rows of vegetables and salad plants, and described how they reuse everything to create a regenerative farming ecosystem. Behind the crops and towards the hedgerow, they keep chickens and a couple of Kunekune pigs. These are tiny pigs originally kept by the Māori in Aotearoa (New Zealand). The animals are not for meat production; they are for companionship and they help to cycle the farm's nutrients. They use the manure from these animals to keep the soils fertile and full of organic matter, preventing the need for synthetic fertilisers. One of Chris and Emily's primary aims is to look after the soils and their microbial communities. In turn, the microbes will keep the soils and the plants healthy.

The pigs are great 'waste' recyclers, and the free-range chickens keep some of the pests, such as slugs, at bay. Chris described how the chicken coop is built on stilts over the top of the compost heap. It has a mesh bottom separating it from the heap. As the chickens do their business, the excrement falls straight into the compost. This is an ingenious idea. The acidic manure helps break down the compost and adds to the organic matter content. The microbes in the compost also break it down and make the nutrients available to plants, and the bugs in the compost provide a nice snack for the chickens – a win–win. In addition, the energy that the microbes and bugs release from the compost as heat keeps the chicken coop warm in the winter.

Chris tells me about some fascinating ecological phenomena that occur on the farm. At one point, they had an over-abundance of moles tunnelling up and knocking courgette plants out of the ground, day and night. 'The courgette plants are an important source of income, and are worth £100 in produce. If the moles damage three plants each night, it soon adds up to a sizeable sum of money. So, I can see why some farmers want to trap and kill them.' He goes on to remark: 'We tried every possible humane way to stop the moles, but they kept on coming, day after day.' Eventually, though, something miraculous happened. One evening, Chris was out walking his dogs on the farm when, suddenly, two tawny owls swooped down with their huge talons and carried away the moles. They were so close to Chris that he nearly hit the ground in true Hollywood war movie style. The owls stuck around the farm for about a week, and the moles haven't been an issue since. This demonstrates that given time and a chance, nature can yield a solution to disequilibrium – often caused by humans in the first instance. Just as the chickens can keep the slugs at bay, the owls can keep the moles at bay, and the slugs and moles sustain the bird populations in return. As the insects pollinate the plants, the microbes keep the ecosystem healthy from the ground up. But these things can only occur if we allow our minds to embrace ecological and regenerative thinking. We must work *with* nature instead of constantly trying to overpower it, and farms like Fanfield are leading this line of thinking in the UK.

There are a few fundamental principles in regenerative agriculture. The first principle is to use the least amount of mechanics to ensure minimal tillage. This helps to prevent soil erosion and allows the soil to build structure. Tillage also creates another disequilibrium by destroying fungal communities and networks. Minimal tillage helps to restore those networks.

The next principle is to armour the soil by covering it with vegetative litter. This enhances the habitat for soil biology, including microbes and the larger critters such as worms and beetles. The soil community can then cycle the nutrients better whilst building a clumpy structure that holds greater amounts of water. Soil armour also mitigates high temperatures and protects against erosion.

The third principle is diversity. Nature abhors a monoculture. A regenerative farm is managed for maximum diversity of plants, animals and microbes. Supporting a diverse crop with grasses and forbs helps feed the soil microbes. Plants and their associated invisible friends can convert nutrients into a useable form for other organisms. Yet different plants and microbes mineralise different nutrients. Therefore, as much diversity as possible is needed to reap the true benefits. Supporting a diverse soil microbiome enhances the overall functionality and integrity of the ecosystem, with cascading impacts throughout the food web.

The fourth principle is to leave roots in the ground for as long as possible. This means using 'cover crops' all year round. This allows the plants to capture carbon from the atmosphere and feed it to the microbes in the soil. Maximising the time a living root can interact with the rhizosphere and its microbes enables a more complex and resilient soil community to develop.

The final principle of regenerative agriculture is to integrate livestock. Animals function as walking composters and seed dispersers. They bring microbiology and fertility back to the degraded soils. Chris and Emily's pigs and chickens fulfil this principle. They also make wonderful companions. Chris said 'regenerative farmers need to embrace delayed gratification in addition to these core principles'. It can take several years before you notice significant changes in the degraded soil. Still, Chris said you sometimes get a nice shot of instant gratification too – each time you plant a seed knowing you are forming a bond with the land, each time you harvest an organic and sustainable crop, and each time it hits you that you're an integral part of the biotic community. Regenerative farming can be extremely rewarding.

As my time at Fanfield draws to an end, Chris tells me about the concept of 'ghost acres'. This term was first used in the 1960s to refer to land abroad that was and often still is used to grow animal feed for the livestock in this country.[12] The same thing can apply internally, for instance if we import compost or animal feed from elsewhere within the country. This could positively impact the farm by providing the resources required to keep the system healthy, but it could harm the land of origin whilst it stays 'out of sight, out of mind'. Moreover, by importing landscaping materials and microbes from other environments, the local ecosystem could change in ways we don't yet fully understand. For example, introducing microbes that are not adapted to the local ecosystem could be detrimental to local plant–microbe relationships. And some opportunists may damage the integrity of the microbial community, which could impact the rest of the system. These issues do not chime strongly with the regenerative farming philosophy, but practitioners in this realm are mindful of them. Indeed, regenerative agriculture prides itself on holistic thinking, and farmers like Chris and Emily are continuously trying to improve their practices for the benefit of humans and the rest of nature – including our invisible friends.

Restoration

I mentioned Dr Martin Breed in Chapter 6. He is a world-leading restoration scientist, who grew up in the suburbs of Adelaide in South Australia. Some of his fondest childhood memories were exploring the little patch of the urban forest in his backyard. He'd make tree houses and underground shelters with his brothers and friends, climb trees and hide from his parents, exploring life in and around their tiny wildlife haven. He spent weeks turning over rocks and logs looking for frogs and skinks, and nights in the tree house, listening to the owls and other nocturnal sounds. He thinks this instilled a strong sense of connection with nature from a very early age.

Martin's PhD looked at the effects of habitat fragmentation and climate change on the ability to find wild native plant seeds for revegetation (i.e. seeking seeds to replant an area following deforestation, for example). He also started thinking about the potential connections people have with the natural world, and whether it was possible to be a scientist with expertise in restoration genomics and work at the interface between people – just another species – and nature.

It's now important to define 'restoration ecology'. This is the science and practice of restoring ecosystems following degradation – for example, after we chop down a woodland or heavily pollute a lake. Essentially, it is humans trying their best to heal or repair damaged habitats.

Martin spends most of his time thinking about two phenomena in restoration ecology. The first is how we can improve our methods of restoring ecosystems to achieve better environmental outcomes. The second is to understand the direct ways in which degraded ecosystems damage human health – and how restored ecosystems can support human health. An emerging theme in restoration ecology is repairing the environment for the purpose of health and well-being. The science of restoration ecology is well established and is a primary toolbox for healing damaged ecosystems. Yet Martin spotted that something was lacking – knowledge that directly connected ecosystem restoration to human health, particularly in urban areas. Martin says that, unfortunately, there is never enough money or interest in repairing the environment for the environment's sake. However, suppose we connect it to human health, to which it is of course intrinsic. In that case, it will only take a fraction of the health budget to repair huge environmental areas and improve human health by default. Many people, governments and companies are self-serving; so, Martin has put a great deal of effort into understanding how the restoration of ecosystems can be good for them whilst trying to shift their mindset to be more attuned with nature.

I asked Martin why and how microbes might be relevant to the field of restoration ecology. He said, to begin with, every plant and animal – including humans – is directly reliant upon and colonised by microbes, and the plants and animals oblige by providing them with habitats and resources. We also know that microbial communities change when people impact ecosystems, but they can also recover when the ecosystems are restored. And when you visit an ecosystem, you get bombarded and are colonised by the microbes that live in the soil, float around in the air, and settle on the plants and animals. Martin and his team are working on ways to restore ecosystems in order to supply humans and other species with a dose of microbes that improve our health. Restoring microbes to restore health is now a cutting-edge pathway linking ecosystem restoration with human well-being outcomes.

Martin and his team have undertaken several ecosystem restoration studies that involve microbes. Understanding how microbial communities change between damaged ecosystems and restored ecosystems can offer new insights into the nature–human health relationship. Few studies have compared the microbiomes of natural versus human-altered environments. Additionally, there is little knowledge regarding which microbes might represent ecosystem restoration – that is, which microbes indicate whether a damaged ecosystem has been recovered towards a more natural, biodiverse state. Dr Craig Liddicoat (also based in Adelaide), Martin and colleagues set out to understand this phenomenon, asking whether we can identify microbial indicators of ecosystem restoration.[13] They collected soil samples from a ten-year eucalyptus woodland 'restoration chronosequence' at Mount Bold, South Australia. A restoration chronosequence is a set of ecological sites that share similar attributes but represent different ages. This allows ecologists to study plant and soil communities that change in a relatively predictive, linear manner over time.

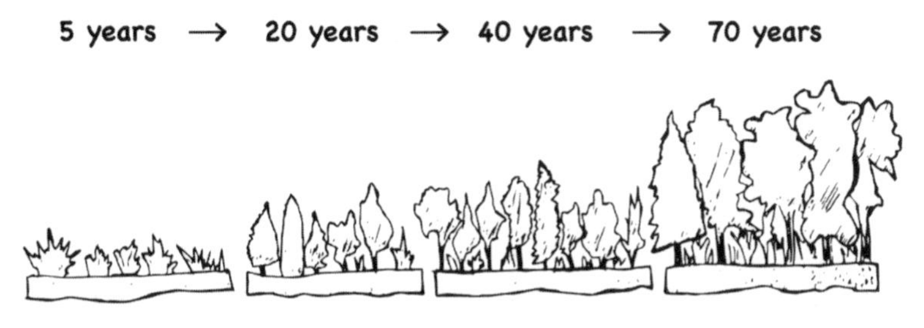

Restoration chronosequence.

Mount Bold is a nature reserve and reservoir about 45 minutes south of Adelaide. The reservoir is surrounded by more than 5,500 hectares of gum forest and is home to more than 160 native animals, including the southern brown bandicoot and the iconic koala. The area has suffered from degradation, including deforestation (mass felling of trees), but it has also benefited from restoration over the past decade. This made it an ideal study site.

Craig and Martin took the soil samples to the lab, where the bacterial DNA was extracted, amplified and sequenced. They then processed the results using a rich bounty of analytical tools. Once the analysis was complete, they found two vital microbial indicators of ecosystem restoration. The first was a group of opportunist microbes that decreased in abundance following ecosystem restoration. This was a fantastic result – especially from a human health perspective. It suggested that if we restore the nature around us – the soils, plants and animals – we'll be less likely to encounter those opportunist pathogens that cause disease. The second microbial indicator was a group of more stable and specialised microbes, or niche-adapted microbes. It turned out that these microbes increased when the natural habitat was restored over those ten years. Amazing.

We can expect restoration to bring increased stability and diversity of plants, which will increase the abundance and diversity of animals that feed on the plants, and then the predators that feed on these plant-eating animals. We might also expect to see a rise in interactions between the plants and the soil, along with the resident microbes. Restored soils would accumulate organic matter, including plant debris. Mycelial networks would re-establish their marvellously complex matrix of hyphae and reach out across the ground, connecting the plants along the way. The soil microbiome would likely increase in complexity and, as a result, shift away from an unstable community characteristic of highly disturbed land.

In disturbed land such as a recently deforested habitat (e.g. for timber production or land conversion), microbial communities and their feedstock (the organic matter) may undergo large fluctuations, and the system could ultimately collapse. At this point, ultra-fast-growing and adaptable opportunist microbes are more likely to thrive as vacant niches become available to them. When restoration occurs, specialist microbes begin to form stronger connections, robust interactions and resilient communities. Some become keystone species, whereby many other species depend upon their presence in the ecosystem. As this ecological complexity builds up, the more fleeting opportunists become outcompeted. This is one of the reasons why as 'niche-adapted microbes' increase, opportunist pathogens decrease. Imagine living in a tiny village where ten people were villains and ten people were the pillars of society. You could soon imagine the mayhem that would creep into town as the villains conspired and carried out their nefarious acts. On the other hand, if the village had ten villains but a thousand diverse pillars of society with different strengths and minds, the villains would soon be caught out, and their impact on the village would be far less damaging.

I asked Martin, 'What do you consider the most important finding of your research to date?' He said it related to a combination of health and biodiversity research. In a mouse study published in 2020, also led by Craig, the team showed that mice exposed to only minuscule amounts of dust from biodiverse soils had lower rates of anxiety-like behaviours.[14] This was connected to the abundance of a particular bacterium in the guts of the mice, called *Kineothrix alysoides*. This is a spore-forming, butyrate-producing bacterium. As mentioned earlier in the book, butyrate is a short-chain fatty acid essential for gut and potentially brain health. In their study, these microbes were most abundant in biodiverse soils, and they had already shown that it was possible to restore this type of soil microbiome. Therefore, the team demonstrated that we can restore a health-promoting aspect of microbial biodiversity, and that this directly connect to mental health. In the future, Martin hopes to carry out more studies, closing this mechanistic link between restoring microbial diversity and human health.

I wondered where Martin might draw his inspiration from now. He describes his research group at Flinders University and says, 'I work with a wonderful suite of motivated, creative and dedicated research students and staff. They're going to make such a positive difference to the world, and that makes me proud.' He also teaches undergraduate students who are creative and passionate about improving the world. This provides him with a tremendous sense of optimism for the future. Martin says the planet has all sorts of environmental issues, but he's impressed that the next generation – at least, the ones he spends time teaching – is up for the challenge. Martin recently became a dad. He is surprised at the sense of inspiration this provides him. 'My little boy (Ollie) is less than a year old now, and thinking about my opportunity to help make the world better for him provides a tremendous sense of clarity for my inspiration.'

Looking forward, I ask what Martin's future research directions are. He says we need to improve how we take care of the environment. Many people on the planet live in highly urbanised areas, more so than ever before. In Australia, around 83% of people already live in urban areas. If we don't take care of the environment, then the health of all creatures, humans included, will suffer. The loss of connection between humans and nature, which we often call the 'extinction of experience', is compounded by urbanisation. Indeed, a recent study showed that rural dwellers connect significantly more to nature psychologically and have far greater knowledge and experience of nature than urban dwellers. This is one reason we need to enhance the biodiversity in our cities. From a microbiome perspective, we need to understand how to optimise restoration strategies so that nature can do its thing and heal us. This is what Martin will focus on, now and in the future.

My chat with Martin was drawing to a close. I asked him a final question: 'What can microbes teach us about the world?'

Martin describes how there isn't an ecosystem on earth that hasn't been impacted by people or microbes. He thinks that it is microbes' rapid ability to change that humans need to be most aware of. He says, 'We should challenge ourselves to think like a community of microbes and see ourselves as part of an interconnected web of life. If we see ourselves as just another species that is connected to the rest of life, it's quite a simple step to see why we must take care of nature and all species to take care of ourselves.' This reminds me of learning about the traditional ecological knowledge of Indigenous Peoples. During my PhD, I worked with some awe-inspiring Indigenous scholars from Turtle Island (aka North America) and the land of the Kaurna Peoples in Australia (aka Adelaide). This taught me a great deal about other ways of generating knowledge, and other value systems, which the Western science machine has oppressed over the years. Many Indigenous cultures view nature as a densely tangled web of interconnected subjects instead of a discrete set of objects as per Western paradigms. We could learn a tremendous amount about living in harmony with nature by ethically engaging with

Indigenous Peoples and embracing these holistic ways of knowing, being and valuing. This will help us rewild our environments, bodies and minds.

A final note in this chapter: have you ever thought about how the light and noise pollution that we humans create, such as from cars, buildings and roads, could affect the human and environmental microbiome?

In the final year of my PhD, this very question sprang to mind. Indeed, human-generated noise and artificial light pollution have increased to alarming levels across the globe. Evidence suggests they can impact ecosystems and human health. However, scientists have given limited focus to the effects of noise and light pollution on microbiomes. As microbial communities are integral to our ecosystems, the effect of noise and artificial light upon them could have important ecological and human health implications. Dr Ross Cameron, Dr Brenda Parker and I reviewed the relevant studies on this topic. We found evidence that human-created noise and light pollution could significantly influence ecosystems and human health by disrupting the microbial communities that live out in the natural world and inside our bodies.[15]

For context, hazardous noise to humans is considered to be anything above 85 decibels (dB), and a lawnmower or a revving motorcycle emits around 90 dB. Importantly, these noise levels can affect how microbes grow and communicate. Suppose it turns out that the noise and light we generate in our bustling metropolises affect our communities of invisible friends to the point where ecosystems are damaged. In that case, restoration ecologists have a new realm to consider. In particular, our urban environments are characterised by a tapestry of street and building lights, along with noise from revving engines and giant machines. To conclude, we proposed the *photo-sonic restoration hypothesis*: restoring natural light and sound levels would help restore microbiomes and, therefore, ecosystem integrity.

The Indigenous botanist and author Professor Robin Wall Kimmerer has written extensively on how being naturalised to a given environment means to live as if the surrounding ecosystems sustain us, and to treat the land as if our future generations depend on it. She says, 'Like other mindful practices, ecological restoration can be viewed as an act of reciprocity in which humans exercise their caregiving responsibility for the ecosystems that sustain them. We restore the land, and the land restores us.' I highly recommend reading her books *Braiding Sweetgrass* and *Gathering Moss*.[16] Robin presents some powerful lessons in these books and wraps them up eloquently in Indigenous science, Western science and poetry.

CHAPTER 11

Bio-Integrated Design

'You never change things by fighting the existing reality. To change something, build a new model that makes the existing model obsolete.'
—R. Buckminster Fuller

The *Anthropocene*. The current geological epoch of unprecedented urban development. Indeed, 70% of the world's population will be living in cities by 2050.[1] Consequently, there is a tremendous urgency to enhance the quality of urban living environments for both humans and wildlife. The 'greening' of urban streets by planting trees and creating parks has been an essential response to our nature-depleted industrial past and present. Nonetheless, this greening typically comprises small green pockets dotted across the dense urban jungle. The exteriors of buildings or the 'architectural skins' make up a disproportionately large surface area in our towns and cities. For instance, there are approximately 44,000 premises in London greater than 18 metres high.[2] If we conservatively estimate the width of each building to be 10 metres, this means the buildings take up 4.4 million square metres – a huge mass of land that was once a rich tapestry of natural habitats for many species. However, we can include the surface area of the walls and windows in our calculations too. In this case, the buildings have created over 30 million square metres of new hard surfaces. Given these extraordinary numbers, it is no wonder the urban fabric is now a prime target for innovative architects who are saying *no* to desolate surfaces and *yes* to building materials that support tiny life-forms of all shapes and sizes.

During my PhD, I was lucky to work with the Bio-Integrated Design lab in the Bartlett School of Architecture at University College London. The lab is headed up by Dr Brenda Parker and Professor Marcos Cruz. Brenda, aka 'algae grrrl', has a background in biology, studying algae and evolution. Marcos comes from an architectural background. He reconsiders architecture's biological, social, aesthetic and technological dimensions. Together Brenda and Marcos have curated a wonderful course where biology and design unite to be greater than the sum of their parts. Their research transcends the boundaries of current knowledge to develop innovative design solutions that we greatly need in the ever-challenging urban environments we live in.

Architecture is beginning to absorb, integrate and host nature in its skin. Indeed, architecture can now be responsive to human activities and their by-products. The air is packed with microbes, spores and seeds. We can now design the architectural skin to be biologically receptive – that is, welcoming to microbes, plants and other organisms – and no longer a bleak emptiness, like the building surfaces of the past.

Marcos gave a talk at the Bartlett, UCL about living architecture.[3] In this lecture, he refers to a project by one of his students, Maria Knutsson-Hall. She designed a building structure that could host algae and cyanobacteria. Maria was inspired by the sloth – the sluggish and inordinately cute Central and South American tree-dwelling mammal. The sloth's fur doubles as an edible garden ecosystem that hosts a collection of diverse microbes, many of which are unique to the sloth. The hairs of the sloth's fur possess cracks and grooves that have the remarkable capacity to hold and retain moisture. This creates a hydroponic (aqueous solution) habitat for algae to be cultivated. Incidentally, the algae and cyanobacteria are a vital food source for sloths, acting as an on-the-go pantry. The algae also serve as an effective camouflage hiding their appearance and odour. The green fur blends in with the trees, and the odour means they also smell like the trees – a perfect strategy for hiding! Taking inspiration from natural features such as sloths and their fur and replicating them in human-centric objects like buildings is known as *biomimicry*. Integrating these features into designs and materials along with other biological considerations is known as *bio-integrated design*.

Marcos is interested in organisms called cryptogams. These include species such as algae, lichens, mosses and ferns. In urban areas, we have largely neglected cryptogams. This is because they are small, and the politics of greening cities have focused on what is greatly visible – parks and trees. However, Marcos says cryptogams account for 7% of the carbon and 50% of the nitrogen fixation from the air.[4] From this perspective, the role of cryptogams in the world's chemical cycling is profound. Marcos's PhD focused mainly on 'architectural skins'. He studied the textures, the contours, the elements, and how these phenomena interact and facilitate colonisation by biological entities. The concept is known as *bioreceptivity*. Academics like Marcos study how biologically receptive a material is – for example, how life-forms colonise it and what factors affect this receptivity.

A bio-integrated material is challenging to achieve because designers need to understand the behaviours and quirks of the organisms whilst also having a solid understanding of the scaffold that hosts them. The designers experiment with a range of materials to create the structural scaffolds for life to take hold. Architects can produce carefully controlled designs with intricate grooves to establish the porosity and 'micro' climates needed by microbes and tiny plants. Marcos's team created wall panels called 'bioreceptive façades'. They were obsessed with the idea that the geometry of the surface mattered when it came to life colonising the panels. After all, geometry guides the run-off of water; it creates shading, it creates different microclimates and ecological conditions, along with varying rates of evaporation. The team uses cutting-edge computers and 3D-printing robots to convert their designs into tangible and functional materials. Super high-tech! Watching a playback of the computer simulations is quite mesmerising. Intertwining branches grow into kaleidoscopic patterns with great intricacy. It is like watching the sped-up formation of ocean corals.

What's even more fascinating is that these intricate patterns and magically complex processes are continuously happening all around us (and indeed, inside us) in the natural world. The budding

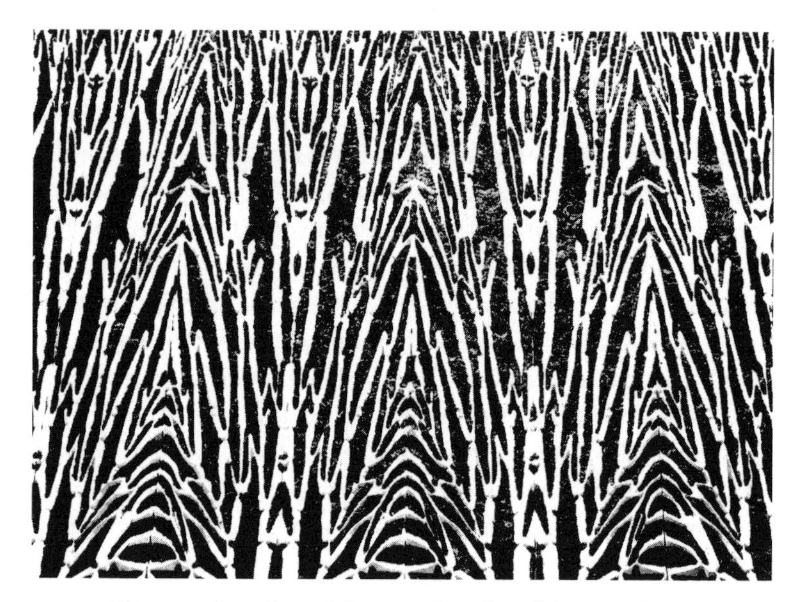

A bioreceptive wall panel design with its (literally) groovy layout!

and multiplication of bacterial comrades into dense biofilm conglomerates; the wiry microscopic branches of mycelium reaching out, searching, and interconnecting beneath our feet; the rich matrix of our human architectural skin dividing and remodelling as new cells and vessels are needed; the expansion of plant tissue cells as the branches and leaves reach high towards the sun. Nature is everywhere and everything. It can teach us great things, if only we allow our minds to be bioreceptive (and bio-respectful).

I went to visit one of the Bio-ID lab's installations in London. It was in a school playground and was designed to act as a natural pollution barrier. It was a drizzly day. We were hoping for a dry spell to collect microbiome samples, to see what invisible friends might have colonised the panels. The installation was a row of bioreceptive panels designed by researcher Alex Lăcătușu, with swirling geometry and peaks and troughs, to provide microclimates for different cryptogams and microbes. Mosses filled the intricate channels, forming a wondrous living (at least temporarily) wall.

After collecting microbiome samples, we headed off to a cafe in peaceful Bonnington Square. This place became famous in the

1980s when all the houses, vacant and awaiting demolition, were subsequently occupied by squatters. The squatters eventually established a volunteer-run vegan cafe, a community garden, a bar and a wholefood shop, and formed a housing cooperative. The cafe and garden remain to this day. Ironically, as we discussed the need to integrate biology into the desolate hard surfaces of urban areas, we spotted a group of fervent volunteers digging moss and 'weeds' out of the nooks and crannies of the paving and keeping the street tidy. Different perceptions of nature and tidiness will pose a considerable challenge to accepting bioreceptive architecture, as we were acutely reminded in that moment.

Algae might not be the first thing you think of if someone asks you to name a microbe. Yet many algae are indeed microscopic organisms. Green algae, for example, are single-celled organisms containing chlorophyll that form dense colonies. Much like plants, algae's chlorophyll provides their often-vivid green appearance.

Scientists now regularly use algae in bio-integrated design. For example, a few years ago, researchers at the University of Cambridge designed a table that could be used as a surface to facilitate photosynthesis.[5] Researchers at Stanford University had discovered that tiny electrical currents could be extracted from algae cells when they undergo photosynthesis.[6] These electrical currents could be used to power various appliances. The Cambridge researchers created the table with an integrated desktop lamp powered by the algal cells!

On a larger scale, architects began thinking about creating entire buildings where they covered the façades with algal tubes, rather than being cladded with steel or glass. This would mean the buildings could potentially produce renewable biofuels, and we could reduce the carbon footprint of the building. In 2013, the world's first algae-powered building was built in Hamburg, Germany, and was called the 'Bio Intelligent Quotient Building'.[7] The Bio-IQ building is a 15-storey apartment block wrapped in dense algae biomass. The algae technology, or 'photo bioreactor', supplies the building with electricity. In addition to electricity, the panels provide heat production, noise reduction, shade and insulation.

The Bio-IQ building has inspired further research into the power of algae-based architecture on larger scales. Look out for more algae-powered buildings and infrastructure in the near future.

More recently, designers created algae-powered breathing pavilions to help combat the effects of air pollution in cities. Air quality is a challenge that many cities currently endure. According to some reports, a child born in Warsaw in Poland today will inhale air pollution equivalent to smoking 1,000 cigarettes in the first year of life, due to the burning of fossil fuels. Other reports put Krakow in Poland at 4,000 cigarettes per year. On a bad day, air pollution in Beijing would be equivalent to smoking 25 cigarettes per day. In Shenyang, where the worst ever levels of air pollution were recorded, this figure is 63 cigarettes per day.[8] More emphasis should be placed on the right to clean air.

To reduce the impacts of city air pollution, the algae-powered breathing pavilions produce breathable oxygen whilst purifying the local air. This is no doubt an extremely innovative creation; however, in my mind, reducing the source of air pollution has to be the priority. Fast and reactive fixes are rarely sustainable, although the breathing pavilions may provide a valuable solution in the short term.

Brenda has worked with algae for many years now. She and her colleagues have created some mind-blowing bio-integrated structures. Brenda considers bio-integrated design to define a pathway towards a future human habitat, where nature plays a central role. This goes beyond simply being a model or providing inspiration and instead is the medium of a multi-layered design approach that is integrated biologically and socially. Her work takes potentially life-changing phenomena such as biotechnology and the impacts of climate change as a foundation to a biorecep-tive and more sustainable built environment. Brenda and her team once created robotically extruded algae gels implanted into ceramic façades. These algae biogels produced new living membranes that absorbed carbon from the air and released oxygen. The algae façades can also absorb contaminants from water. Designers have also developed ceramic tiles covered with protective layers of bacterial biofilms that self-assemble to create wonderful glowing light (bioluminescence). You've probably seen the wildlife TV programmes where someone brushes their hand in the ocean at night, and a flurry of vivid blue and white light is emitted. This is bioluminescence caused by chemical reactions in marine microbes.

Brenda explains that cyanobacteria also convert carbon dioxide (a gaseous substance) into calcium carbonate (a solid substance). This mineralisation provides a material with structural integrity for applications in architecture. Amazing. So microbes can also make solid materials that can be used in our buildings.

Brenda also describes how food waste can be processed and reconfigured into complex structures, allowing for a truly circular-economy approach in architecture. Additionally, biological solar panels can convert light energy into electrical energy using the photosynthetic capacity of microbes. We can install these microbe solar panels on roofs and inside buildings to reduce energy consumption powered by fossil fuels. It seems the world of bio-integrated design is booming and is ripe for innovation.

I supervised a Bio-ID lab master's student during the summer of 2021, namely, Will Scott. Will is highly innovative and certainly has a bioreceptive mind. The title of his thesis was 'Creating prebiotic materials to introduce exposure routes to health-promoting microbes in the built environment'. Will recognised that the effects of changing lifestyle norms on the microbial ecology of the built environment (i.e. our towns and cities) were relatively underexplored. Will was also acutely aware that our environment and our health interconnect through the microbial denizens around us and inside us. Could we design architectural materials to host and enhance health-promoting bacteria? And could the materials be produced in a sustainable and circular-economy manner? These are the questions Will and I grappled with during his thesis project. It has to be said that I was merely an armchair supervisor due to COVID-19 restrictions at the time. But we had some delightful and thought-provoking chats via video calls.

Will would send me regular updates as he experimented with different 'biomaterials'. He tried using rice, soya, glycerine, coconut coir and even eggshells, mixing them to create a bioreceptive scaffold for the bacteria. There were so many things to consider when making the material. Could it retain moisture? Did it have the correct pH to support the bacteria? Could it provide the proper nutrients for the bacteria to feed on? How long would the material last before degrading? How could we make the material durable enough to be helpful in an architectural context?

We investigated the different types of culture growth medium preferred by the bacteria, to optimise the nutrients in the materials. Many years of experimental fine-tuning have allowed scientists to determine the optimal growth media for various species of bacteria – each having their preferences for food and micro-ecological conditions. Way back in the 1860s, the first bacterium was cultured in a reproducible way. This was done by the French microbiologist Louis Pasteur.[9] He fashioned a medium out of a yeasty soup containing ash, candy sugar and ammonium salts. He noticed that the chemical features of the soup could promote or impede the growth of different microbes and that competition for the nutrients between microbes occurred. This resulted in some bacterial species outcompeting others and dominating the culture medium. Robert Koch subsequently discovered that broths based on meat extracts enhanced the growth of various bacteria. Scientists later referred to Koch as the 'Father of Culture Media' to reflect his groundbreaking findings. Later, Koch experimented with agar after Fanny Hesse, the wife of Koch's research assistant, suggested it.[10] She had been inspired by the use of agar in the preparation of jams and fruit conserves, a method that had been employed for hundreds of years in different parts of Asia. Fanny learnt of this practice as a child in New York, where she lived next door to a Dutch-Javan immigrant who used to make jam! Agar proved to be a superior general-purpose gelling agent, and it is still widely used today.

The *Streptomyces* bacterium Will wanted to use preferred a soy-based medium, and the *Bacillus* bacterium preferred a rice-based medium. Will also noticed that the *Bacillus* had been isolated on rice many times before and so decided to incorporate rice into his biomaterial. This also meant that Will could try to source his ingredients from food waste products, thereby providing that all-important sustainability factor. Will came up with an excellent name for his finished products – 'prebiotic materials'. Will's experiments revealed that prebiotic materials could sustain the growth of his selected bacteria for a period of at least ten days. For comparison, Will tested other materials frequently used in the built environment (e.g. ceramic, steel and wood) and found that they failed to sustain growth for the same period. This was a fantastic finding so early on in Will's career. Although we need more research, prebiotic materials could represent a viable approach to introducing health-promoting microbes into our cityscapes.

Richard Beckett, an academic at UCL, also questions what architecture should look like in the age of the Anthropocene.[11] He considers how buildings and cities can be better integrated as a biologically productive element of the biosphere. He describes our current urban fabric as often dead and inert, and explores how to amalgamate living matter into buildings to encourage a resilient and healthy future.

Richard has designed wall tiles embedded or 'biologically seeded' with living bacteria that can inhibit the growth of the superbug MRSA. This is a novel approach to creating indoor surfaces for buildings. Surfaces inoculated with living bacterial communities could alter the indoor microbiome and promote a healthier living environment. I think Richard and Will need to join forces – sustainable prebiotic and probiotic materials unite! In addition to fostering beneficial microbes, the materials could potentially help reduce the pathogenic types whilst limiting the spread of antibiotic resistance. The bacterium Richard used was *Bacillus subtilis*. A recent study showed that *B. subtilis* could kill *Staphylococcus aureus* – the bacterium of MRSA[12] – in both planktonic (free-floating) and biofilm form. Incredibly, *B. subtilis* also increased *S. aureus*'s susceptibility to antibiotics such as penicillin and gentamicin. The researchers went a step further and tested the effectiveness of *B. subtilis* at reducing the negative impacts of *S. aureus* in a mouse model. They found that it did indeed reduce the *S. aureus*'s presence and increased its susceptibility to other antibiotics like penicillin. This is a very promising area of research.

Scanning electron microscope images showing the probiotic wall material porosity at 10× magnification (left), and the *Bacillus subtilis* cells dispersed within the material matrix as a biofilm at 2,500× magnification (right). Images provided by Richard Beckett.

This next story is about electricity, and eventually microbes, of course! In 1752, Benjamin Franklin conducted his famous experiment with a kite, a key and a lightning storm. He attached a wire to the top of the kite to act as a lightning rod. He secured a hemp string to the bottom of the kite, which conducted an electrical charge. The silk string he held in a dry barn deliberately prevented electrical conduction. He attached the metal key to the hemp string. During the storm, the loose threads on the hemp string started to stand erect, and when Franklin moved his finger towards the key, he felt a spark as the positive charges in his hand attracted the negative charges in the metal. He eventually used a Leyden jar to store the high-voltage electrical charge. This was one of the first recorded demonstrations of the connection between lightning and electricity. However, it was not the discovery of electricity, as many accounts document. Over 2,000 years before this event, the Ancient Greeks had discovered that rubbing fur on amber caused an electrical attraction.[13] This is the first recorded discovery of static electricity. In the early twentieth century, archaeologists discovered pots with an iron rod and sheets of copper inside. Although the evidence is ropey, some believe this was an ancient battery that produced light in Roman times.[14] By the seventeenth century, several electricity-related discoveries had been made, including an early electrostatic generator. Later in 1800, Alessandro Volta constructed an early battery – the voltaic pile – that produced a steady electric current, and in 1831

Michael Faraday created the electric dynamo, which enabled him to produce a constant electrical current. And with a blink of an eye in evolutionary timescales, we now have satellites, fMRI scanners, laptops, mobile phones, aeroplanes, automobiles and the internet.

Therefore, it is only relatively recently that humans have used electricity. You may then be surprised to learn that several common bacteria have been using – and even generating – electricity for millions of years. Enter *Geobacter*, the mud-dwelling bacteria!

Professor Derek Lovley, at the University of Massachusetts Amherst, is the head of the Geobacter Project.[15] Derek and his team found that some bacteria naturally produce electricity. They discovered that *Geobacter* evolved the ability to 'breathe' solid lumps of iron in the soil. By doing this, they can generate electricity. Humans can harvest this electricity from the bacteria using devices known as microbial fuel cells. The fuel cells have two electrodes. One is an anode, the other is a cathode, and the two link together by an electrical connection.

As we know, many organisms, including humans, inhale oxygen and exhale carbon dioxide. The *Geobacter* bacteria have a different strategy because they evolved without lungs and in environments that lack oxygen. The *Geobacter* ingest the organic matter in their muddy environments and 'breathe' out electrons produced as part of their normal metabolism. The *Geobacter* 'breathe' these electrons onto the anode in a microbial fuel cell. Scientists can then collect the electrons as an electrical current. To complete the electrical circuit, the electrons pass to the cathode. And boom, a biological battery is formed. At any given moment, trillions of these *Geobacter* bacteria are buzzing and exchanging electrons beneath our feet.

The 'exhaled' electrons need somewhere to go in the natural world. They typically go into an element like iron oxide, which is bountiful in soil. The *Geobacter* have an unconventional method of exhaling these electrons. They use tiny filaments called *protein nanowires*. These structures are thousands of times smaller than the width of a single human hair. Using powerful microscopes, Derek and his team could view them. The nanowires can transport electrons thousands of times the length of the *Geobacter*, despite their diminutive size. A nearly unimaginable feat. For instance, try holding your hand out in front of your mouth and then taking a normal breath in and out. Feel how subtly the exhaled gas molecules caress your hand. Now imagine being able to breathe out and instantly shuttle those gas molecules over a mile into the next village, across the countryside, or to the other side of your city. This is how impressive *Geobacter* respiration is – arguably the most impressive of any known organism on Earth. They evolved this adaptive trait in response to the extreme conditions they face deep underground – though the conditions are only 'extreme' from a human perspective; they are normal for the *Geobacter* and their adaptive traits. *Geobacter* can also use their protein nanowires to communicate with other microbes such as archaea.

Derek and his team, including nanowire specialist Professor Nikhil Malvankar, found that *Geobacter* grown in the lab have another neat trick up their sleeve. When stimulated with a small electrode, the *Geobacter* assemble into dense biofilms. They stack up like high-rise apartments and move electrons through a single shared electric grid. One question vexing the team was, how can the *Geobacter* on the top floor of the biofilm apartment block shuttle electrons, sending them to the bottom of the pile and out through a nanowire? Now that the bacteria were in biofilm formation, the distances were tremendous.

The team went to work. They used state-of-the-art microscopy techniques; in fact, they used two types of microscopy.[16] One method they used is called *infrared nanospectroscopy*. This allowed the researchers to identify unique amino acid fingerprints in the nanowires by analysing how they scattered infrared light. The other technique is *atomic force microscopy*. This allowed them to gain high-resolution information about the nanowire structure by using ultra-sensitive mechanical probes to touch the nanowire surfaces. Imagine reading Braille, but the raised dots are one ten-millionth of a centimetre small. Mind-blowingly high resolution.

The researchers found that when an electric field stimulated *Geobacter* in the lab, they produced a special metallic nanowire. This was 1,000 times more efficient than *Geobacter*'s nanowires in their natural habitat – the dense and sticky mud. This allowed

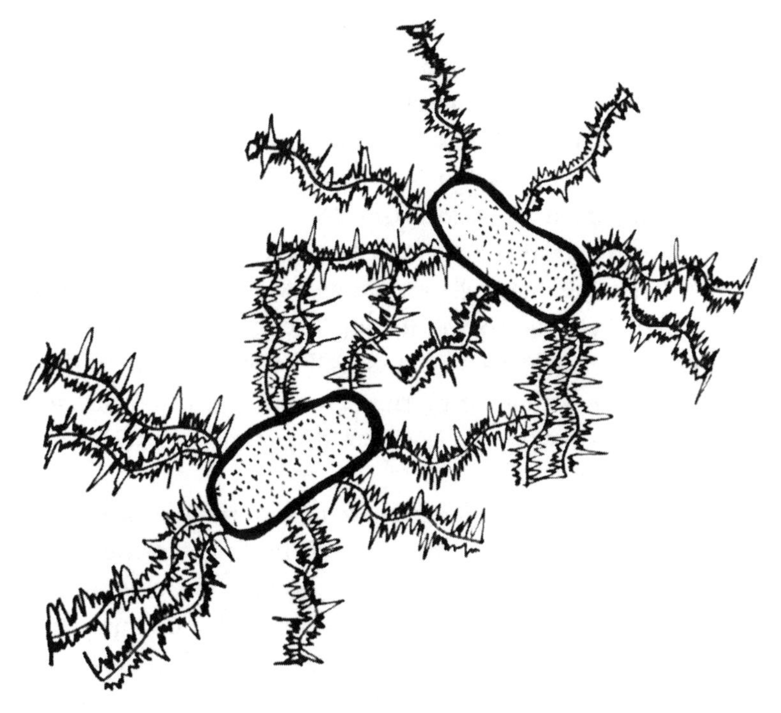

Electrogenic *Geobacter* with their nanowires.

them to fire electrons across a tremendous distance, thousands of times the length of their bodies!

Researchers like Derek and his colleagues have grown *Geobacter* colonies to power electronic devices for over a decade. A considerable benefit of these microbial fuel cells is their ultra-longevity. I recall watching a video with Derek talking about *Geobacter*. He gives a tour of his lab and points to the star of the show, the 'old man'. This *Geobacter* grew on an electrode for over 12 years, constantly generating electricity and powering contraptions like calculators and watches. Derek says they are interested in the nanowires and their ability to conduct electricity because of a massive worldwide problem. The problem is electronic waste. One report suggests we produced 53.6 million tonnes of electronic waste, such as computers and mobile phones, in 2020; this is equivalent to 200 Eiffel Towers in weight.[17] This is primarily the result of urbanisation, industrialisation and higher disposable income levels. Derek says nanowires are the ideal components for sustainable and disposable electronics. They are renewable with no toxic components and can even be rewired into the microbes to power themselves. I was not surprised to learn that the military have experimented with *Geobacter*'s electrical capabilities. In 2008, they powered a weather buoy in a river for more than nine months with a *Geobacter* fuel cell. According to Derek, the nanowires can currently produce power for small-scale medical monitoring devices like wearable patches. He says the nanowires can function in living tissue without triggering an adverse reaction. In 2006, taxonomists named one *Geobacter* (*G. lovleyi*) after Derek to recognise his work.

Derek, Nikhil and colleagues are now working to strengthen the nanowires to conduct more electricity. This, they hope, will usher in a whole new generation of eco-friendly, microbe-powered batteries. We know fossil fuels are destroying life on our planet, and many organisations promote electric vehicle batteries as the next best and sustainable alternative. However, even these generate a whole host of nasty by-products, destroy habitats and displace Indigenous communities in the process. Could our invisible *Geobacter* friends help us make genuinely sustainable batteries on a massive scale? I guess only time will tell.

Thinking back to Will's bio-integrated design project, I recall another student in his cohort, Agathe Chevee, experimenting with *Geobacter* to power biologically responsive architecture. Agathe used electricity-generating bacteria as sensors to recognise changes in environmental conditions, such as acidity, air quality or alterations in temperature. This would then trigger an electronic response, and tiny bulbs would emit light of different intensities to reflect the amount of a chemical compound, heat or acidity in the environment. This responsive architecture could help future-proof our urban areas by allowing them to adapt to environmental stress. Biosensing is ever-present in most organisms, including humans. For example, the human skin senses environmental stimuli and adapts to protect us from harm. When it's cold, tiny subcutaneous muscles contract, our skin hair rises (called piloerection), and our blood is shuttled in towards our core to regulate our body's temperature. As Agathe put it, even if we only use responsive architecture in a visual sense, it can inform the community of the environmental state at any given moment, in particular the air quality. This can create meaningful connections between humans and our surrounding materials; it fosters the notion of materials as 'interfaces to visualise the invisible forces of nature'. Imagine buildings lighting up when air pollution was above a certain threshold.

Interestingly, researchers are using *Geobacter* and many other bacteria along with other microbes for bioremediation. This is the process of using microbes to clean up soil, air and water contaminated by discharged chemicals. The bioremediation process stimulates the growth of microbes that use the contaminants as a food source – yum! The microbes degrade the contaminants by gobbling up the chemicals, making the environment cleaner and safer. Designers are starting to integrate bioremediation systems into their architectural projects. For example, I mentioned the algae gels that produce new living membranes to absorb carbon from the air and release oxygen. And algae façades can absorb contaminants from water too.

Designed landscape features called *rain gardens* can also provide a form of bioremediation. Rain gardens are shallow landscaped depressions filled with nature-based features such as a mixture of plants and beneficial microbes. They capture, clean and absorb stormwater run-off from roofs and roads. Once the water and pollution have been absorbed, bacteria and other microbes get to work and break down the pollutants, often into less ecologically harmful forms. Rain gardens can also provide pockets of natural habitat for wildlife, and often rate highly on the aesthetics scorecard by integrating features such as diverse wildflower blooms. In addition, researchers are designing architectural materials with filters that can entrap metal ions from industrial discharge. They can embed the filters with bacteria that gobble up the metal ions, making the environment much cleaner. As with my earlier comment about air pollution, these bioremediation systems are incredibly innovative, but ideally, we should be focusing on reducing the pollution at the source.

Failing to mention fungi and slime moulds in bio-integrated design would be remiss. Fungi, in particular, are integrated widely in architectural materials, whereas slime moulds are mainly used for design inspiration. In 2014, a strange building began to take shape in New York City.[18] In its early stages, you would have been forgiven for mistaking it for an igloo. But eventually, the ice-white bricks rose into monumental towers. It was captivating to see. But the building's appearance was arguably not its most impressive attribute. Indeed, the building had been cultivated using 10,000 bricks made of agricultural waste and mycelium. The building and construction industry is responsible for 39% of human CO_2 emissions, and 21% of these come from the production of steel and concrete alone. Using waste that's already degraded allows the fungi to act as a sustainable alternative to standard building materials. Once the mycelia grow into the bricks, they are often heat-treated to prevent further growth as the waste they consume provides structural integrity. Some researchers are experimenting with living fungal materials. These have the benefit of being able to respond to the environment and potentially heal themselves. Imagine living in a house that, if damaged, could heal itself! One idea is to construct walls with two layers of heat-treated fungi enclosing a layer of living fungus inside that can adapt to changes in environmental conditions.

Jonathan Dessi-Olive at Kansas State University in the United States says fungi reshape how we think about the permanency of architecture.[19] What if some of our buildings were meant to only last a couple of years and then be recycled into food or energy? This is a tantalising thought. No doubt we'll see more fungi-integrated designs in the future.

CHAPTER 12

Microbiome-Inspired Green Infrastructure (MIGI)

'If you really want to study evolution, you've got to go outside sometime, because you'll see symbiosis everywhere!'
—Lynn Margulis

Principles of ecology apply at myriad scales, including within the human body and the intertwined visible and invisible ecosystems we depend upon for survival. For instance, predatory species hunt down prey species in our guts, and unwavering cooperation occurs between the microbes in our armpits and in our mouths. These interactions also happen between mammals in a savannah, plants in a forest, or fish in a coral reef ecosystem. The principles of dysbiosis or 'life in distress' apply to different realms of life too. Our gut microbiomes can be 'in distress' and imbalanced if we regularly feast on unhealthy foods, spend time in polluted environments and consume a tonne of antibiotics. This is akin to our woodlands and meadows being 'in distress' if they are disturbed by pollution, felled trees and over-grazing. Parallels are everywhere.

We've already discussed how we can view the human body as a holobiont – a host plus trillions of microbes working symbiotically to form a functioning ecological unit. Well, we can also view the rest of the nature surrounding us as a vast collection of these holobionts. We share the invisible constituents (the microbes) of our human holobiont with all the other holobionts (the plants and animals), and they share their invisible constituents with us. It's an unseen and subconscious gift economy.

I find this notion of interconnectedness and symbiosis inspiring and exciting. By building huge cities and living in them (i.e. the process of urbanisation), aren't we restricting this natural exchange amongst the living community by creating novel environments where nature's collection of holobionts are shut out of our lives? Or where we shut ourselves into these great desolate cocoons? We've spoken about how this extinction of experience and inhibition of natural interactions might affect our health and well-being. We often treat natural features such as trees in our urban areas like furniture – a collection of disposable objects. Cities, by their current definition, are breeding grounds for human-centrism. Trees provide floral and fractal patterns that are aesthetically pleasing to humans. Urban plants do x, y and z for humans. What little urban birdsong is left enhances human well-being. But this focus on instrumental value and only what is good for humans is detrimental to the living community as a whole.

It seems we need a cultural transformation. A profound shift in mindset. From individualism to collectivism. From just humans in urban environments with sporadic natural 'furniture', to a flourishing urban ecosystem that provides for humans and the entire living community. As discussed, microbes have a fundamental role in our ecosystems. This is not to say, though, that it's a 'bottom-up' affair, whereby less complex creatures have an upwards influence on more complex creatures. Indeed, we could argue that a mishmash of upward and downward causation is at play in our ecosystems. After all, cells are not simply just a few organelles floating inside a membrane. A combination of X-ray and high-tech microscopy has shown us that individual cells are a rich jewellery box of life and complexity – a buzzing invisible metropolis of activity. These cells, in turn, form aggregates that have a downward influence on arguably less complex systems in animal and plant bodies. These trigger a flurry of activity and behaviours that impact the integrity of the ecosystem. In essence, our actions affect microbial communities and entire environments, and these microbes and environments affect our actions. It's all connected.

Arguably, the natural next step in following this recognition of interconnectedness is to promote beneficial relationships between the constituents of the whole. The whole being the planet, and the constituents being our environments, our societies, our 'selves', our microbes and our genes.

With all this in mind, I conceptualised *Microbiome-inspired green infrastructure*, or MIGI.[1] This acronym features prominently in this chapter. It is a collective term for the restoration, design and management of living urban features that could potentially enhance the health of all urban dwellers – the humans, birds, trees, bats and bees – and encourage more nature into our cities and our bodies. It specifically focuses on the need to consider microbes when designing and restoring urban environments.

My colleagues and I wrote a preliminary outline of MIGI in my first ever publication.[2] We wanted to emphasise that all forms of life – both the seen and the unseen – are in some way connected, ecologically, socially and evolutionarily. And that this paves the way to valuing reciprocity in the nature–human relationship. This holistic perspective can help tackle the connected human health and ecosystem issues. We aimed to help address the global challenge of halting and reversing dysbiosis in all its manifestations. Ambitious? Sure thing!

As we pointed out in our paper, the holobiont concept or viewing ourselves as communities instead of individuals can be difficult to embrace; consciousness as a biological phenomenon is steeped in intrinsic complexities. It is much easier to view ourselves as an individual of a species. Even as individuals, a fundamental asymmetry exists between how we view ourselves and others, due to our deeply complex emotions and cognition. But once we acknowledge that we are essentially walking communities exchanging invisible life-forms with our environments, we can use ecological principles to help guide our social policy and behaviour.

Our paper explores these ecological principles in greater detail, highlighting the links between human, microbial and environmental health. Landscape and social interventions can enhance our connection with the natural world through interactions with beneficial and diverse microbes. This is where MIGI comes in.

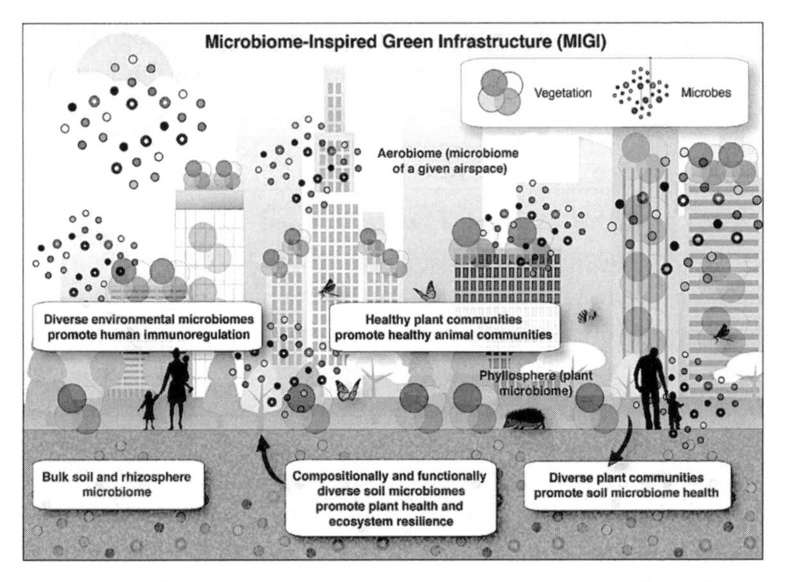

MIGI – healthy urban ecosystems include a diverse consortium of microbes, plants and animals living symbiotically.

The 'inspired' part of MIGI implies a considerable design element. Design considerations might include understanding how pollution and urban features such as buildings, roads and vegetation influence the microbial communities in our cities. It could also involve elements of bio-integrated design – all those wonderful innovations discussed in the previous chapter. MIGI strategies should also aim to maximise ecological justice and reduce health inequalities by ensuring equity of access to health-promoting environments. Healthy urban ecosystems need a diverse consortium of microbes, plants and animals (including humans!) living symbiotically. When this relationship is threatened, the ecosystem fails, and so does our health.

Admittedly, our first paper was still quite human-centred. During my PhD, I endeavoured to develop the MIGI concept to be more inclusive of the wider living community – the plants and animals we share the land with, and the ecosystem as a whole. I ended up splitting the concept into two subcategories – the first focused on human health, and the second on 'ecosystem health'.

Below, I'll discuss snippets from the first paper (still quite human-centred), showing how MIGI can be implemented. After that, I'll discuss how MIGI developed during my time at the University of Sheffield and where I see the concept sitting now.

Foraging

The number of species in the human microbiome has gradually decreased as human populations have passed along the following trajectory: *Foraging → Rural farming → Urban industrial lifestyles.* Some even refer to this as a 'mass extinction event', albeit on a microscopic level.

We know there's a link between a diverse human microbiome and the foraging lifestyle. Therefore, it is envisaged that MIGI will include foraging-friendly green spaces in urban areas. We can learn from our ancestors' lifestyles – progression through regression. To create foraging-friendly spaces in our cities, a collaborative effort between landscape architects, ecologists and urban planners is required. From a MIGI perspective, creating foraging areas could enhance our interactions with environmental microbes and increase the diversity of microbes that live in and on the human body whilst fostering behaviours that benefit the environment.

Foraging also expands the multisensory experience – encouraging us to touch, look and smell. This brings intrinsic advantages, as health benefits arrive through various senses. Despite foraging being globally ubiquitous, it is often discouraged in urban areas. We highlighted that some concerns need to be addressed, such as urban pollutants and the perceived 'mess' from fallen fruits. Nonetheless, urban foraging can take several forms, from harvesting the fruits of street trees and bushes to community gardening in food forests. In our first MIGI paper, we elaborated:

> The former calls for innovation in planting
> design and plant protection along with
> strategies to reduce pollution. The latter would
> benefit from shifting perceptions of the value
> of these food sources, mobilised perhaps

> through community groups such as the *Grow Sheffield's Abundance Project* – an initiative that encourages the harvesting of food plants across the city and redistributing the produce to food banks and local communities.

Urban foraging projects may need to adapt to the complexities of urban life; for instance, the ownership of urban land regularly changes. However, recent innovation is helping to address this issue, as our paper discussed: 'Mobile allotments, such as those created by the *Avant Gardening* project, can be installed on vacant lots to provide communities with a foraging hub and can be easily moved if the land status changes.' They're like vegetable plots on wheels.

Community gardening can promote other benefits, including physical exercise, mental well-being, nature connectedness and social cohesion. So, it's a win all round for human health. We can improve our interactions with our invisible friends through physical engagement with the soils that support the food plants. Allow the soil microbes to connect with your body's bustling jungle of life and exchange microscopic goodies.

As mentioned briefly in Chapter 2, cutting-edge research by Professors Chris Lowry and Graham Rook on the soil bacterium *Mycobacterium vaccae* shows that it can activate serotonin in the brain's prefrontal cortex when injected into mice. This helps to regulate coping responses to 'uncontrollable stress'. And the potential health benefits of *M. vaccae* don't end there; it can protect against brain inflammation and cognitive dysfunction, and has immune-boosting effects in animal models. This is just the story for one soil bacterium. It speaks volumes for the other many billions of bacteria and archaea on the planet, of which, we have only classified a few thousand species.

The possibilities for foraging and the MIGI approach are manifold, and in our paper we wondered: 'Are there specific natural habitats that can optimise interactions with health-promoting microbes? Can we isolate different microbial species and influence communities to enhance these interactions? The research is in its infancy, but the potential is immense.'

Green barriers

We also highlighted that natural green barriers such as hedgerows with trees could be designed in a MIGI intervention. This could help improve people's multisensory experience whilst reducing pollution in green spaces by trapping noxious particles. However, we should scrupulously curate these features to allow wind to disperse some pollutants – otherwise, it could actually increase the pollution concentration in the local area. Additional research is required to understand the impact of green barriers. Still, they could help shield humans and microbial communities from industrial pollutants and reduce cardiovascular maladies. Despite these potential benefits, it is worth noting that these features can also be potent allergen-producers. Yet pollution can mess around with the microbiome of tree pollen grains and enhance their allergenic power, thereby increasing human allergic diseases. As I've said previously, we must tackle the root cause of the pollution in our urban environments; simple fixes are rarely sustainable.

Cultural and ancestral microbiomes

Due to the increasing cultural diversity of Western towns and cities, there's a need to create socially inclusive urban green spaces. Our paper pondered whether we could explore this inclusivity from a microbiome perspective. In theory, creating ancestral environments that consider ethnicity in urban areas could optimise microbial exposure, ensuring equally distributed benefits between different ethnic groups. Some evidence points to differences in human immune responses linked to ancestry, and populations vary in susceptibility to diseases. Moreover, the make-up of the human microbiome differs significantly between different ethnic groups. Factors such as demography and diet only partially explain these differences.

Restoring ancestral microbial communities in natural environments may improve the health of different human populations. It is possible that the historical mixing of human genes across cultures has neutralised this need; however, evidence questions the universality of broad microbiome-based strategies due to ethnic and geographical variation. In other words, one size might not fit all when it comes to microbiome-based therapy.

Ancestry aside, there are cultural and generational considerations for MIGI. These days, many children spend far less time in the great outdoors interacting with nature. This can be due to changes in cultural norms and social barriers that prevent access to green spaces. We discussed how we could design MIGI with cultural changes in mind. Imagine creating microbiome-friendly environments in areas where children spend much of their time. We could design skate parks so that children get a healthy dose of beneficial microbes. Or we could work with developers of augmented reality games such as *Pokémon GO* to promote time in biodiverse environments. After all, this technology is unlikely to disappear. While we can make strong arguments to reduce 'screen time' and encourage contact with nature (a view I fully endorse), this interface between technology, humans and nature could provide new links to encourage physical activity in green spaces. The next step is to understand the potential health-promoting effects of 'nature exposure' whilst using this technology.

The prospect of including different ancestral environments to promote generational and cultural inclusivity is a tantalising one. 'However,' we added, 'additional research is needed to understand the relationships between culture, ancestral environments, microbes, and health across different ethnicities. Moreover, the potential impacts of including novel environmental features in native ecosystems should also [and always] be considered.'

Plant microbiome selection and engineering

Like in humans, interactions between plants and their co-evolved microbes have a role in ensuring plants remain fit and healthy. Microbial diversity can help drive this fitness. MIGI strategies that aim to benefit human health by enhancing microbial diversity could also generate important benefits for plants, and the rest of the creatures in the vast ecological web. This further highlights the importance of the interconnectedness of life.

'Understanding how plant communities affect the microbiome is important,' we stated in our paper, 'particularly in designed urban environments'. We can select powerful interactions in plant communities to improve the soil microbiome and stave off pathogens. It is also possible to select plant microbiomes that enhance the fitness of the plants. Understanding the relationship between our floral and invisible friends will be essential to the success of MIGI. Careful plant-microbe selection processes may have a role in the future design, implementation and impact of MIGI strategies in our novel urban environments.

I mentioned Harry Watkins in Chapter 9. We have a running joke that makes a fleeting but regular appearance when we're chatting about research. Our eyes light up like a Christmas tree and voices become animated when talking about microbes, trees, evolution and urban health. During our PhDs, our conversations, which happened in various places – from traditional pubs in Northern Ireland to ten floors up in Sheffield's Arts Tower building – usually resulted in one of us saying, 'Let's start a research project to investigate this!' The first step in our exciting new research project would be to produce an academic conference poster. We aimed to present our ideas at conferences and generate interest from peers. With all limbs crossed, we then hoped an inquisitive rich person would walk by and throw loads of money our way, thereby enabling us to start the study. Plan B (and back to reality) was to apply for funding from a research body. By year three of the PhD, we had ideas to produce about ten posters – and we presented several of these across the country. That rich person never walked past. So, now, whenever an exciting knowledge gap in microbial ecology comes up, we joke, 'You know what this calls for? It's poster time!' or words to that effect. We haven't officially announced our retirement from mass-producing posters, but I think it's an unspoken truth.

Anyway, still strapped for research funding, we eventually decided to create something that could be impactful, but that didn't require too much cash. With sketch pads and marker pens in hand, we formulated a plan to write a publication. This time, the focus was on how microbial ecology could be overlaid onto established architectural plans and procedures. The Royal Institute for British Architects (RIBA) has a seven-stage 'plan of works'. This provides a framework for urban landscape projects from conception and design to construction. We noticed that none of the seven stages considered microbes. But they could do! Whether it be the inclusion of vegetation and soil parameters to boost microbial diversity or foraging areas that promote direct interactions between humans and our invisible friends, considerations for microbes could be included.

I am pleased to say this paper was published,[3] and it led to some exciting collaborations with people both inside and outside of academia. For example, city planners are now starting to integrate considerations for microbiomes into their urban masterplans. They are designing areas in cities to improve the environmental microbiome with the view of enhancing human and ecosystem health.

We could view MIGI as simply a conceptual framework – a basic structure underlying a series of principles. It aims to promote biodiversity in urban areas, with specific considerations for microbes and their benefits to humans and the rest of nature. The concept is under development, and there is still a great deal to know. Our next publication aimed to discuss precisely what was known about microbial ecology in relation to human and urban ecosystem health, and to scan the horizon to see what could soon be realised in this exciting realm.

Vegetation, microbiomes and the built environment

One important consideration within the field of MIGI is to ensure we expose humans and other species to an abundance of different microbial species from a young age. Air samples downwind from biodiverse vegetation contain more diverse microbial communities than upwind samples. Therefore, a relatively simple MIGI strategy for urban designers could be to create public spaces and buildings downwind from these floriferous sources or, perhaps more importantly, integrate biodiverse vegetation within building structures and spaces – bio-integrated design! This has the dual benefit of boosting habitats for wildlife.

I briefly touched upon this in Chapter 6, but my colleagues and I recently showed that the microbial communities in the air flowing through urban green spaces differed depending on the height of the airspace – with bacterial diversity decreasing with height. This implies that we should increase vertical planting in urban areas, allowing for exposure to higher natural microbial diversity in the vertical dimension; do people living at the top of tall buildings, for instance, receive a less healthy dose of microbes due to the lack of greenery?

Evidence suggests that planting native species can really work wonders for the urban soil microbiome. Diverse native communities of plants on the surface can encourage stronger relationships below ground, and these can increase the resilience of the ecosystem. Just think of a diverse and well-established human community. If the members of the community have different skills, know each other well, accept each other, and help each other in times of need, then the community will be more resilient in the face of adversity. If the community is not inclusive and people only care about themselves, the community as a whole will be fragile. MIGI strategies could include planting diverse, and where possible native, plant communities to boost the health of our urban ecosystems.

However, a possible issue arises when it comes to only planting native species – climate change. Our warming climate may cause issues because native species might not adapt to the changing conditions. As we stated in our 2018 paper, 'It is not yet clear how native plant populations will tolerate future climate conditions. Studies on woody plants offer conflicting views. Some research suggests that local genetic variation may provide sufficient resilience, whilst others argue that given the range in possible climate futures, including species beyond those that are locally native may be vital.'

MIGI strategies should also connect urban and rural habitats by providing natural corridors that spread across the city like a vast green web. Many urban environments are 'patchy' in natural features such as hedgerows, trees and meadows. These are vital for wildlife – a great many birds, invertebrates, mammals, reptiles and amphibians call hedgerows their homes. Such features also strengthen the integrity of the ecosystem and, thus, the microbiome. To ensure our urban ecosystems stay healthy in the long term, we should connect all the natural habitats within towns and cities and provide essential corridors to the broader landscape. From one perspective, our cities are like enormous parasites on the land. Yet we can change this by improving species interactions and resilience across the landscape. Creating urban habitats with diverse networks of vegetation can help animals, plants and microbes connect. A world without connections is a world without meaning, hope and life.

Soil microbiomes

The properties of soil have a tremendous influence on environmental microbiomes. The soil's organic matter and clay content are associated with structure, nutrients and water-holding properties. Therefore, soil properties create the conditions for microbes to flourish. A vital question for MIGI strategies will be 'Should we use in-situ or imported soils?' We would expect some soil types (e.g. loam – a healthy balance of sand, silt and clay) to promote higher microbial diversity. 'Sandy soils provide suboptimal microbial habitat, while heavier clay soils may be prone to poorer drainage in wet climates or plant stress in dry climates,' says colleague and soil scientist Dr Craig Liddicoat.

Foot and vehicle traffic can also compact and degrade the soil. This could adversely affect the soil microbiome. Imagine if a giant alien stomped all over your house, crushing the structure and annihilating the food and water. You're not going to fare well.

Moreover, plant communities require complex mycorrhizal networks, acting as conduits of plant communication whilst fighting pathogens and promoting adaptation, growth and memory. Therefore, we must consider the condition of the soil and its role in sustaining our ecosystems. Interestingly, landscaping materials (e.g. compost and mulch) have a 'microbial shelf life' much like the cereal in your cupboards or the milk in your fridge, and storing them for long periods (which often happens) can reduce the beneficial microbes and the life-giving nutrients. This suggests that

MIGI strategies should consider encouraging a short-term storage time for landscaping materials.

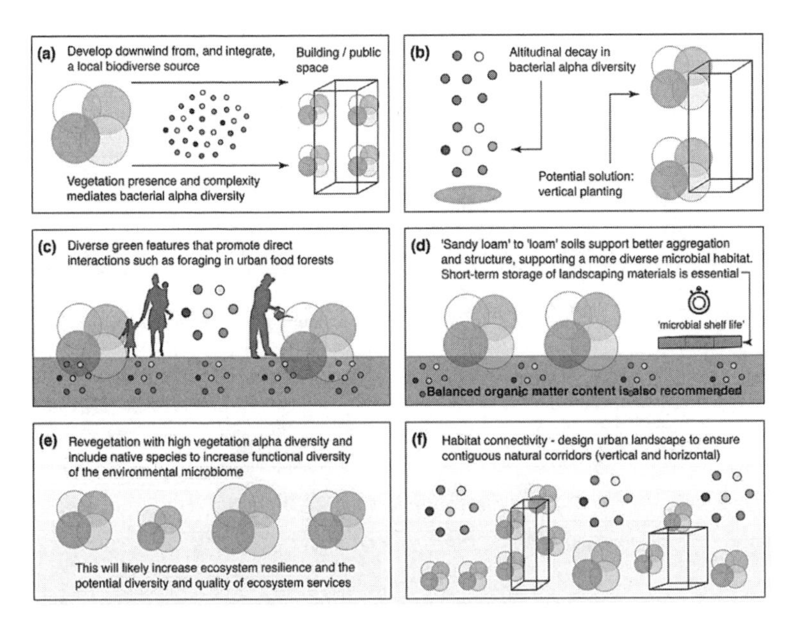

MIGI – current possibilities.

Horizon scan of near-future possibilities

I spoke about bio-integrated design in the previous chapter. Well, this sits within the MIGI framework too. Creating architectural skins that integrate biology to enhance human health and urban biodiversity is central to this approach. I didn't mention the concept of microbial inoculants in the previous chapter. Researchers have recently used inoculants to shift the microbial communities in landscaping materials, such as park benches and sandpits for children, towards a community that protects the immune system. A wonderful Finnish research group recently developed a microbial inoculant from biodiverse sources such as forest materials. They inoculated the sandpits in children's play areas.[4] After the children had contact with the inoculated sand, the number of opportunistic pathogens on their skin significantly decreased. This aligns with Marja Roslund's excellent biodiversity intervention study I spoke

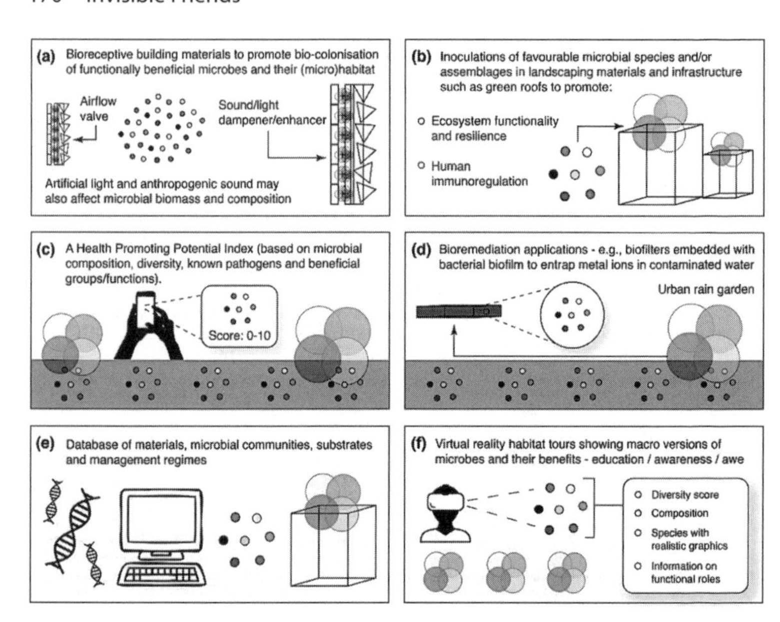

Near-future applications of MIGI.

about in Chapter 2, where researchers brought forest biodiversity into school playgrounds to enhance the children's microbiomes and, thus, their immune systems.

Other researchers have shown that microbial inoculants can also benefit plant health through the power of plant-growth-promoting bacteria. These bacteria can protect plants from stress caused by drought and heavy metals, and as their name suggests, they play essential roles in plant growth. They could help reduce the use of harmful synthetic chemical fertilisers. MIGI strategies could therefore incorporate microbial inoculants to enhance urban ecosystem health.

Researchers are developing useful tools to help MIGI interventions. For example, someone recently produced a tool to help assess the environmental microbiome's influence on plant traits. People can use this tool to monitor and enhance plant growth. I am currently developing a tool so that urban developers can optimise the microbial components of their environments for both human and ecosystem health. A smartphone app could soon allow users to scan an area of habitat and determine whether it might be good or bad for their health based on its microbial residents. Virtual reality systems can also facilitate urban habitat tours. This could include interactive displays of microbial communities whilst providing information on our invisible friends' mighty roles in the local ecosystem.

As our species – or our collective holobionts – moves forward in the Anthropocene, we must address the global challenge of halting and reversing dysbiosis in all its manifestations. Perhaps an epoch that we could aspire to is the *Symbiocene*.[5] This is a term first coined by Glenn Albrecht, the Australian 'eco-philosopher', highlighting the need to take a more holistic, mutualistic and ecological approach to the way we live.

There are countless possibilities for MIGI, particularly if we work collaboratively to progress the concept. Imagination is vital to the equation. As Albert Einstein purportedly said, 'Imagination is more important than knowledge. For knowledge is limited, whereas imagination embraces the entire world, stimulating progress, giving birth to evolution.' There is an opportunity for a concerted effort to create urban environments that are conducive to life. Life for humans, but also life for our non-human friends. Imagine a future where this is a reality, and it can be so.

CHAPTER 13

To Catch a Thief: Forensic Microbiology

'Everything is a portrait. A diary. Your whole drug history in a strand of your hair. Your fingernails. The forensic details. The lining of your stomach is a document. The calluses on your hand tell all your secrets. Your teeth give you away. Your accent. The wrinkles around your mouth and eyes. Everything you do shows your hand.'

—Chuck Palahniuk

It was late 1987. A Floridian girl, Kimberly Bergalis, had just turned 19 years old. Many would assume she had a long life ahead of her. At the time, Kimberly was in good health but had issues with her teeth. She booked an appointment to see her dentist, Dr David Acer.[1] He suggested that she have two molars removed to fix the problem. Kimberly agreed, and David removed her molars. The tooth removal procedure was seemingly successful. The dentist was paid, and Kimberly went home.

At the time he saw Kimberly, David must have been ill for many months. Earlier in 1987, doctors diagnosed David with acquired immune deficiency syndrome. AIDS can manifest as several life-threatening infections and illnesses when the human immunodeficiency virus (HIV) adversely affects the human immune system. Kimberly started to feel sick in March 1989 while she was a business major at the University of Florida. Although she became extremely ill, she was unaware of the cause. Later that year,

Kimberly nearly died of pneumonia. The first symptoms included a persistent sore throat and tonsils, and a fungal infection in the mouth. Rapid weight loss followed. It wasn't until January 1990 that Kimberly was diagnosed as being HIV positive. The tardy diagnosis was perhaps unsurprising. At the time, Bergalis was a strict Catholic who had boyfriends but had never been, according to her, sexually active. She had no history of blood transfusions or drug use. Her doctors tested for many diseases, including diabetes, hepatitis and leukaemia. The results all came back negative. They decided to test for HIV, and the results were shocking; she had AIDS. Following her diagnosis, I can only imagine Kimberly and her family feeling utterly bewildered, trying to figure out how a healthy girl who was not sexually active and did not take drugs could have possibly acquired HIV. The only medical procedure she had had in the previous couple of years was the removal of two molars at the dental surgery. Perhaps, in the back of their minds, there was a slight inclination to suspect this as a route of infection. It was a matter of months before her diagnosis was officially linked to the dentist, who died of AIDS later in 1990. David wrote to all his patients just days before he passed away to inform them he had AIDS. Too little, too late.

Over 600 patients came forward and were tested for HIV. Four others tested positive, including a retired English teacher who, like Bergalis, had no risk factors for HIV. Government health officials sequenced the viral DNA from both Kimberly and David. The results? The two virus isolates were almost identical. The Centre for Disease Control concluded that the two HIV isolates were closer than any other two ever assessed in North America at the time. Somehow, while extracting her molars four years prior, David had infected Kimberly, along with, it would seem, several other patients. Scientists reportedly estimated the probability of transmitting HIV in this situation to be 0.3 in 1,000. Kimberly was the first person in the United States to be infected with HIV by a healthcare worker, so this type of transmission was a rare occurrence and one that took cutting-edge forensic microbiology to uncover. We had only invented polymerase chain reaction or 'PCR' – the process used to make billions of copies of a DNA fragment – a few years prior. For over 100 years, microbiology played a relatively small role in forensic science, until advances such as PCR led to a boom in forensic science applications. Kimberly's case was tragic, but despite her fading body and mind, she fought until her final days to ensure measures were in place to prevent this type of transmission from occurring again. Sadly, Kimberly died of AIDS in late 1991.[1]

The technology used in Kimberly's case (of an invisible foe) helped spur on forensic microbiology more widely, and microbes have since provided a helpful hand in the investigation of criminal cases. In the mid-1990s, forensic researchers developed a process to analyse fungal and pollen spores. This allowed investigators to differentiate between soil types, which enabled them to link contaminated items to the soil of a given environment. Despite the technological breakthroughs of the 1990s, it was not until the early 2000s and the rise of bioterrorism (terrorism using biological weapons such as anthrax) that the forensic microbiology field emerged in response to these growing threats. Historically, forensic microbiology was constrained by a lack of available and cost-effective DNA sequencing technologies. However, improved technology and data processing techniques will likely move forensic microbiology into the next generation of crime solving within the next decade.

As of March 2019, the conviction rate for homicides in England and Wales in the UK was 79%.[2] That's 21% of homicides that did not end in a conviction. This is slightly higher than in the United States, where the conviction rate is around 70%.[3] An astoundingly high prevalence of wrongful convictions also occurs around the world. Often, insufficient evidence is available to convict a perpetrator of a crime. The Innocence Project is a national litigation and public policy organisation dedicated to exonerating wrongfully convicted individuals.[4] According to them, 375 people in the United States have been exonerated by DNA testing, including 21 who spent considerable time on death row.[5] For every eight people executed in the United States since 1973, one person has been wrongfully convicted and later exonerated.[6] Consequently, there is a genuine interest in augmenting the forensic toolkit, including cutting-edge microbiology and molecular methods.

As we've discussed, microbes are abundant in and on the human body (microbial cells can outnumber the total number of human somatic cells), in surrounding environments, and on objects associated with a crime. A growing body of research suggests that investigators could use forensically relevant profiles of microbial communities as evidence or, at the very least, microbes could complement traditional forensic methods. The field is still in its infancy, but the potential of understanding microbes to help 'catch a thief' is immense.

In the following sections, I give specific examples of where microbes have been or could be used to help criminal investigations. The potential applications of microbiomes (i.e. communities of microbes) still require further development, but there are a number of exciting applications in the making. Lovers of popular crime dramas, read on. And those of you who aren't – beware, I do mention corpses a couple of times.

Geolocation

Researchers have carried out intensive work in the past few years to find out which invisible creatures reside in our towns and cities. These studies have demonstrated that unique microbial communities exist across a city along with 'molecular echoes' of environmental events. For example, people leave behind their DNA, their human cells and their microbes on surfaces. Investigators could potentially use these molecular echoes as evidence to link a person to a given location.

Forensic scientists recognise the potential of analysing microbial communities in the soil. The make-up of the soil microbiome is unique from location to location, differing in the ground beneath your feet, for example, and a spot 400 metres away. Therefore, microbiome samples could be helpful in locating the origin of soil when soil is found on an item linked to a crime – for instance, the sole of shoe. One study demonstrated that we could elucidate the origin of the soil obtained from the sole of a shoe by comparing the similarity of microbes on the shoe to different soil samples.[7]

Researchers have also investigated seasonal changes in soil microbes living in corpses.[8] They found that after death, microbial communities changed in specific and reproducible ways (like natural succession on a microscopic scale). Investigators could use the daily changes in the body's microbiome in the soil to estimate the time of death. This could provide a vital piece of information in a crime scene investigation.

Unlike the natural succession that occurs when a grassland naturally converts to a forest on a timescale of years, changes in the microbial communities within a dead body occur daily. For example, scientists have observed the rapid growth of Firmicutes bacteria from the bloat stage (early decay) to the advanced decay stage (yuck!). In one study, these Firmicute bacteria at first made up less than 5% of the relative abundance of microbes living in and around the dead body; whereas by day 12 their abundance increased to 75%.[9] The work is ongoing, but it points to a promising tool for estimating how long a body has been in a location after death (known as post-mortem interval). Microbes could therefore help crime scene investigators determine the time of a person's death.

Researchers can now pinpoint the location of a city in the world, with an accuracy of close to 90%, by analysing the microbiome of office surfaces.[10] This means that offices have city-specific

Soil and dead body (cadaver) microbes might hold forensic clues.

microbiomes. Your office in London will have distinct microbial signatures to someone else's office in Melbourne in Australia, or São Paulo in Brazil. Even within countries, the office microbiome will be distinct to the city – so, the Melbourne one will likely be very different to the Sydney one, and the same for Mombasa and Nairobi. Many of us spend over 90% of our time in built environments such as offices and homes. It may soon be possible to determine which city you are from by analysing the microbes on your body, clothing or belongings. Another important factor to consider is that skin microbiomes differ between humans living at high and low altitudes. One study collected skin microbiome samples in Tibet and found consistent enrichments of several bacterial species in samples collected from higher altitudes.[11] All of this suggests a potential route to help determine the origin of a body.

Personal identification

Each one of us humanoids may be uniquely identified based on the microbial communities living in and on our bodies – our microbial 'fingerprints'. This could have a considerable impact on forensic science, especially if the investigator is unable to retrieve sufficient amounts of human DNA at a crime scene, which is often the case. Yet, it is unknown whether the variation in microbial communities between people is sufficient to identify individuals within large populations or if the microbial communities are stable enough across time. In one study, gut microbiome samples were used to pinpoint 80% of individuals up to a year later.[12] Another study had similar results from skin microbiome samples. The results are encouraging, but researchers still need to improve the accuracy if these methods are to be trusted in a court of law.

Trace evidence

We leave swathes of microbes behind on objects and surfaces – also known as 'trace evidence'. For instance, huge numbers of bacteria cover personal objects such as mobile phones, wallets and glasses; many of these microbes come from the owner's body. Moreover, human items such as shoes and mobile phones can support distinct

Microbes could help with personal identification.

microbial communities – that is, they differ consistently between individuals. One study demonstrated that bacterial communities sampled from mobile phones were significantly more similar to their owners' microbes than the microbes of other people, which is probably to be expected.[13] Another study found that post-mortem skin microbiomes could be linked to personal objects with a high level of accuracy.[14] Several of the items in the study were associated with 100% accuracy. These included spectacles, bottles and steering wheels. Computer devices were associated with around 70% accuracy. This suggests that we might be able to connect skin microbiome samples with objects at crime scenes in the near future. So, remember, even if your human DNA is not left behind, your microbes might be!

Manner of death

An expert determines a person's manner of death following an investigation. The expert may be a coroner, the police or a medical examiner. Five different 'manners of death' are considered in forensics: natural, accidental, suicide, homicide and undetermined.

Trace evidence: you leave signature microbes behind on items you touch.

Scientists recently found that different microbial species were linked with the manner of death.[15] In this study, Xanthomonadaceae bacteria were more prevalent in hospital-related deaths, whereas *Actinomyces* species were more prevalent in suicide cases. The accuracy of these microbial tools still needs to be improved, but researchers are combining microbial ecology with sophisticated computers to establish more reliable 'biomarkers' – biological markers or 'indicators'. Think about the canaries that miners used to carry into the coal mines. If levels of noxious gases such as carbon monoxide became too high, the gases would kill the canary before killing the miners, thereby providing a warning to exit the mine immediately. Biomarkers or bioindicators work in a similar way (although not necessarily as early-warning systems) but on a microscopic level – they provide indications of different phenomena such as the time of death or the condition of the body. Another example of bioindicators in the visible world are plants. Take the common nettle as an example. This pioneering hairy plant of temperate climates prefers fertile and often disturbed soil. Therefore, it provides an aboveground visual indication of the subterranean conditions. In the Northern Hemisphere, dense carpets of moss often grow on the north side of trees due the cooler and damper conditions there. You can use this as a bioindicator of cardinal direction if you find yourself lost in the woods without a compass. Many bioindicators help tell a story. By joining forces with microbes, forensic scientists could soon piece together the story of a person's death.

Cause of death

The term 'cause of death' means something different to 'manner of death'. 'Cause' refers to the disease and injury that produces damage in the body and leads to death. For instance, investigators can use microbes to help determine cause of death by drowning. Drowning is one of the leading causes of unnatural deaths worldwide. Researchers have analysed the presence of tiny algae called diatoms for well over a decade now. This method has been the 'gold standard' for determining death by drowning; however, researchers question its reliability.

Several studies have supported death by drowning diagnoses by detecting the DNA of bacterial species associated with aquatic environments, such as *Aeromonas* in ponds or lakes.[16] Bioluminescent (light-emitting) bacteria such as *Vibrio fischeri* and *Vibrio harveyi* have been used to determine the cause of death by drowning in seawater. Another study analysed the microbes of both drowned and post-mortem rats, comparing freshwater versus marine-water treatments.[17] They found that many bacterial species were unique to either the marine or freshwater ecosystems, thereby allowing forensic scientists to tell the environments apart.

Animal microbiomes

Several studies have shown that different animals possess unique microbial signatures. For example, chickens, macaques and plateau sheep have unique gut microbiomes.[18] These microbiomes are shaped by genetic and geographical factors. Additionally, the skin microbiomes of finches, amphibians, bats, cetaceans and dogs are known to be unique. Interestingly, humans share microbial communities with their dogs – we live with them and exchange thousands of invisible particles with them daily. Some of these particles include microbes. And some of these microbes may colonise our bodies or the bodies of our animals, thereby creating a potential forensic connection between humans and pets.

Additional research is required, but there is potential to link a person with a location based on their shared microbes with animals. Microbial profiles obtained from other species could be associated with a given environment or occupation – for example wild animal industries, pet industries or pet ownership. You may think you've gotten away with a crime, but your pets (or their microbes) might say otherwise.

Indeed, scientists could potentially detect animal microbes on a suspect or victim's body, clothing or belongings. These microbes may be helpful when human or non-human animal DNA evidence is lacking. Investigators could trace microbial profiles to the point of contact with an animal, or an environment such as equine stables, pet shops, zoos, or even a pet-dwelling household. This approach is primarily theoretical, but this type of microbial profiling could complement other traditional forensic evidence in the future.

Future opportunities and challenges

It is currently uncommon to use microbiome-based forensics in criminal investigations and the court of law. Nonetheless, suppose sufficient research is invested in ascertaining the microbial characteristics of the environment across space and time. In that case, reconstructing the context linked to a criminal act becomes relatively straightforward. For example, imagine an investigator finds a corpse buried at a depth of two metres in an arid environment in the summertime. And the temporal succession of the cadaver's microbial community for these environmental conditions has previously been established. In that case, we could infer the time of death with a high degree of accuracy. *Reconstruction*, in this sense, is probably the more accessible category when it comes to using microbiome science in forensics. The other potential category is *comparison* – for example, comparing the microbial signature on an item of interest with the microbes of a perpetrator or victim. Comparison tools may benefit the criminal investigation but are harder to develop with a high level of accuracy. As the forensic scientist Dr Zohar Pasternak said in a paper we recently wrote together: 'The most beneficial way to employ such tools would be in a "*one-to-many*" configuration, similar to forensic human DNA analysis, whereby a DNA profile from trace evidence is compared to all the profiles from known persons and locations in a database.'[18]

By taking a DNA profile from, for example, a murder scene (i.e. 'the one') and comparing this to a DNA profile in a database (i.e. 'the many'), the forensic investigator can check for matches. If they find a match, they can calculate the probability of encountering this profile by chance – that is, the chance of it originating from a location or person unrelated to the crime.[19]

There is a profound need to build relevant forensic databases of microorganisms that we can compare to trace evidence. Currently, Pasternak says the only way to proceed in a forensic context is in a 'one-to-one' configuration. For instance, investigators could compare microbiome samples from a suspect's shoe or cell phone to samples found at the crime scene for each criminal case. This approach can provide relative conclusions, such as one sample being *more similar* to another sample. However, we still need an extensive microbiome database and detailed ecological knowledge regarding the factors that shape microbiomes. This will allow for conclusions such as the probability that one sample (on an item such as a weapon) is the *same* as another sample (on a human); and calculations of the chance of it being from another human – for example, 1 in 100 million.

The Phantom of Heilbronn

Perhaps one of the most challenging hurdles to overcome is the issue of DNA transfer. In the past couple of decades, human DNA evidence has gained widespread credibility in the court of law. In fact, we have rarely disputed DNA trace evidence connecting a person to a crime scene until recently. It is becoming more and more common for the defending party to challenge DNA evidence. The defence might now suggest that the DNA was deposited at the crime scene by a legitimate activity – for example, when the suspect held an object or shook the victim's hand, or, of course, the hand of the actual perpetrator. As microbiome data is more dynamic than human DNA data, we will need considerably more research before the legal system accepts microbiome data on a regular basis. However, the potential for microbiomes to complement traditional forensic methods is promising.

Many other challenges remain in forensic microbiology, such as contamination and ethical concerns. Notably, even well-developed human DNA-based evidence is not immune to errors.

The Phantom of Heilbronn case exemplifies this issue.[20] We often read fictional stories about criminal masterminds who evade detection time and time again. But these masterminds are not always confined to the realms of fiction. The Phantom of Heilbronn, often referred to as the 'Woman Without a Face', was a hypothesised female serial killer in the 1990s and 2000s. Investigators tied six murders and various other crimes to the Phantom. For years, the Phantom ran rings around the European law enforcement agencies, puzzling the most proficient detectives. The killer's existence was inferred solely from a molecular calling card – the DNA discovered at several crime scenes in Germany, Austria and France between 1993 and 2009.

It was April 2007, and two police officers took a break in their patrol vehicle in Heilbronn, Germany. A young police officer, Michèle Kiesewetter, 22, was behind the wheel, and her partner, Martin Arnold, was in the passenger seat. Chatting away and smoking cigarettes, the officers were approached by at least two people. The officers probably thought they would be asked for directions or help, so they rolled down their windows. However, they weren't asking for directions. Instead, the approachers shot both police officers in the head. Sadly, Kiesewetter died at the scene, and Arnold fell into a coma. The perpetrators reached into the car and stole the officers' weapons and ammunition, and fled the scene.

Miraculously, the gunshot to the side of Arnold's head wasn't fatal. Weeks later, when he awoke from his coma, he had little recollection of what happened. He couldn't recall what the perpetrators looked like or how many were present. Luckily for the crime scene investigators, one of the killers had left a trace of their DNA in the patrol car. Presumably, the perpetrator who reached into the car to steal the weapons had deposited some skin cells. The DNA analysis revealed that the suspect was probably female and of Eastern European descent. Perhaps the most critical finding was that the killer was already in the police system. They had struck before – several times, dating back to the 1990s. This was the Phantom of Heilbronn.

The (wo)manhunt ended up consuming 16,000 hours of police overtime and millions of euros in expenses. This immense effort centred around creating a profile for one of the world's most wanted criminals. Rather than going into hiding after the Heilbronn media storm, the Phantom seemingly carried on murdering people in Europe. In 2009, the police investigated a corpse found in a burnt-out building back in 2002. The forensics team were chasing a new lead. They attempted to match the DNA from the corpse to that found on a 1990s asylum application form. Incredibly, they found that the DNA on the form matched the Phantom! The lab technicians stopped to think, thank goodness. Why on Earth was the DNA from someone with completely different sex chromosomes and ethnicity showing up in their results? They repeated the tests with a different swabbing method, and the Phantom's DNA was nowhere to be found. Had the Phantom crept into the lab and switched the results? Of course not. This anomalous conclusion was the undoing of the entire fiasco. The police could now track the Phantom down to her hideout – at a German packaging company. The packaging company contracted with a major cotton swab supplier, who employed several Eastern European packers. By now, you may have already guessed who the Phantom is (or isn't); if so, you'll be far ahead of the European police forces, who took several years to work it out!

Slowly but surely, the debacle was unravelled. The Phantom was not a single perpetrator. It was a forensic blunder. Forty-three crimes in various countries were pinned on this single illusory serial killer, and the DNA was from contaminated swabs. One factory worker, who was packing the swabs, was responsible. She was innocent when it came to the homicides but clearly needed lessons in contamination control. Her DNA had gotten all over the cotton swabs before packing, and the police, unaware of this, were sequencing her DNA after swabbing various crime scenes across Europe. Moreover, it turned out that the cotton swab supplier had never even stated the swabs were suitable for collecting human DNA samples; the police force had probably been using incorrect swabs for many years. Nowadays, the police force only uses swabs decontaminated with ethanol – after all, it would be hard not to learn a lesson from this embarrassment.

In 2011, two neo-Nazi gang members committed suicide together in an apartment in Germany. When the police investigated the apartment, they found Officer Michèle Kiesewetter's gun and her blood on a tracksuit. This evidence suggested that these neo-Nazis were responsible for the shooting of Officer Kiesewetter in her patrol car in Heilbronn. Her killers were likely part of a dark criminal gang who never faced justice.

This story tells us that forensics, especially when using highly sensitive material such as DNA, is not always the slick and reliable process you see on TV shows like *CSI*. This is why the field of forensic microbiology must take heed of these lessons. By doing so, the potential to join forces with microbiomes to solve crimes could be realised in the not-too-distant future.

CHAPTER 14

Microbes in Outer Space

'The cosmos is within us. We are made of star stuff. We are a way for the universe to know itself.'
—Carl Sagan

Panspermia. This is the hypothesis that life exists throughout the known universe. But how are life-forms distributed from one space rock to another? According to this theory, it's dust particles, meteoroids and comets, as well as spacecraft, that inadvertently host a plethora of microbial hitchhikers.[1] The word comes from the Ancient Greek, *pan* for 'all', and *sperma* for 'seed'. To test the hypothesis, astronauts took some hardy bacteria to the International Space Station (ISS) to see how resilient they were in the face of zero gravity, zero oxygen, zero water and zero many-other-things. The bacterium of choice was *Deinococcus radiodurans*.[2] This microbe is listed in the *Guinness Book of World Records* as the 'world's toughest bacterium'. It can survive in places where water is absent and nutrients are few, enduring a thousand times more radiation than a human can. Its name means 'strange kernel that endures radiation'. No other known organism can be exposed to as much radiation as *D. radiodurans* and survive; it's clearly the prime candidate for space experiments.

To withstand these harsh conditions, *D. radiodurans* has an extraordinary trick up its sleeve. The system it uses for repairing DNA is out of this world (pun intended). Exposure to high levels of radiation seemingly decimates its genome. However, the microbe can stitch its fragmented genome back together with absolute fidelity. Interestingly, there isn't a habitat on Earth that exposes life to anywhere near the amount of radiation that *D. radiodurans* can endure. This poses questions as to why it would evolve such resilience in the first instance. Some researchers think the trait evolved as a mechanism to adapt to dehydration, which causes similar types of damage to the genome. Some scientists have suggested that the special configuration of the *D. radiodurans* genome is partially responsible. Unlike many other bacteria, which carry just one copy of their genome, *D. radiodurans* carries between 4 and 10, stacked on top of one another! It also hosts an expanded repertoire of DNA repair proteins that scramble to sew the genome back together with maximum efficiency.

Astronauts took *D. radiodurans* to the ISS in the late 2010s and deposited in pellets on the station's exterior. After one year, researchers monitored the bacteria using electron microscopy tools. Incredibly, *D. radiodurans* did not exhibit any morphological damage (i.e. in form or structure).[3] Space orbit had induced a rearrangement of molecules in the microbe and triggered several stress responses that allowed it to survive the hostile conditions. It could survive when exposed to wavelengths below 200 nm. This is well into the harmful UV end of the electromagnetic spectrum, which ranges from 100 to 400 nm. Scientists consider this important because the UV spectrum of Mars is below 190 nm due to the shielding provided by carbon dioxide. Researchers are pursuing applications of *D. radiodurans* in simulations to help them predict where to search for life on Mars. Perhaps more importantly, some are considering using *D. radiodurans* to help understand how to avoid cross-contamination between life on Earth and other 'alien' life-forms. In my opinion, we must consider the ethics of the deliberate or accidental introduction of microbes to other places in the solar system. Indeed, we can see from our activities on Earth, so-called 'invasive species' can wreak havoc on our ecosystems.

Another intriguing potential use of *D. radiodurans* in space is in sewage treatment on long-haul space flights. The bacterium is known to break down chemicals into less harmful forms. As science journalist Sarah DeWeerdt said, 'The mere existence of *D. radiodurans* suggests that almost anything may be possible.'[4]

The interior of the ISS is a closed system inhabited by a relatively diverse assemblage of microbes arriving from the cargo and crew. We would traditionally use standard culturing methods (by swabbing and growing in Petri dishes) to monitor the microbial communities on the ISS. However, as very few bacteria can be cultured, sequencing methods are required to understand which microbes call space their home. In 2019, a team of researchers sampled the microbiome of the surfaces in the ISS and analysed them upon their return to Earth.[5] The researchers found that many species of bacteria and fungi had colonised the ISS surfaces, including three strains of Methylobacteriaceae that were new to science. This family of bacteria is typically found in soils, where they often help to promote plant health and nitrogen fixation. Scientists think these arrived at the ISS on the astronauts or in their cargo. The three novel strains were closely related to *Methylobacterium indicum*, a bacterium often found on rice seeds. This was an interesting finding; it's important to discover novel microbes that could help promote the growth of plants in hostile conditions whilst being able to endure those conditions themselves, and *M. indicum* could be one such candidate. There are likely many other microbes on the ISS that weren't sampled, so watch this space (pun *not* intended).

In 2018, an advanced closed-loop system was installed on the ISS.[6] This is an air-recycling plant about two metres tall and one metre wide. It is part of the life-support system and generates oxygen for the astronauts. The closed-loop system uses some carbon dioxide from the cabin to convert into methane and water. The water then feeds back into the system to produce oxygen, which minimises the need for water deliveries from Earth. The system can't use all of the carbon dioxide from the astronauts' breath to produce water, so a novel addition was created involving microbes – more specifically, micro-algae. Scientists integrated a photobioreactor on the ISS that hosts *Chlorella vulgaris* – a resilient alga that undergoes photosynthesis to produce oxygen. This increases the efficiency of the closed-loop system, pumping fresh oxygen out to the astronauts. A bonus to this hybrid system is that when fully functional, the algae could double up as a suitable food source.

In principle, this means the ISS and other spacecraft can carry less food, further increasing the sustainability of the missions. It has been suggested that *C. vulgaris* could replace around 30% of the astronauts' food supply due to its high protein content.[7] It is hoped that developing these hybrid systems could also improve sustainability options on Earth by promoting a circular economy.

Whilst working with the Bio-ID lab at UCL, I learnt how researchers are developing 'micro-printed 3D surfaces' for the ISS. These are similar to those structures designed to host microbial life on bioreceptive walls, but they are being trialled to capture excess water particles. Whilst this doesn't involve microbes directly, the inspiration for the materials comes from a biomimicry approach (mimicking natural surfaces that support diverse life-forms) aimed at hosting microbes on architectural materials. In the future, we could see the interior of the ISS plastered with bio-inspired designs and oxygen-pumping walls, thanks to our invisible friends.

A chapter on microbes in outer space would not be complete without discussing tardigrades. Unlike bacteria, they're not unicellular but are, in fact, tiny animals. We often refer to tardigrades as 'water bears' or 'moss piglets' because of their mammal-esque appearance and lumbering movements. And similarly to our bacterial friend *Deinococcus radiodurans*, they are inordinately resilient in the face of extreme stress. The name tardigrade comes from the Latin *tardus* for 'slow' and *gradi* for 'to walk', reflecting their slow-stepping nature. Tardigrades walk on every continent on Earth, including Antarctica. They were first 'discovered' by the German pastor Johann August Ephraim Goeze in 1773.[8] Around 1,300 known species exist on the planet, but given their ability to endure extreme environments, it is likely that many more haven't even been discovered. They are prevalent in organisms like mosses and lichens and feed on plant cells, microbes and tiny invertebrates.

Due to their immense resilience, tardigrades have survived all five of the planet's mass extinction events. They have been reported in hot springs, on top of the Himalayan mountains, and even in the deep-sea abyss. Some species can endure hot temperatures of up to 150°C and cold temperatures of −272°C – that's close to absolute

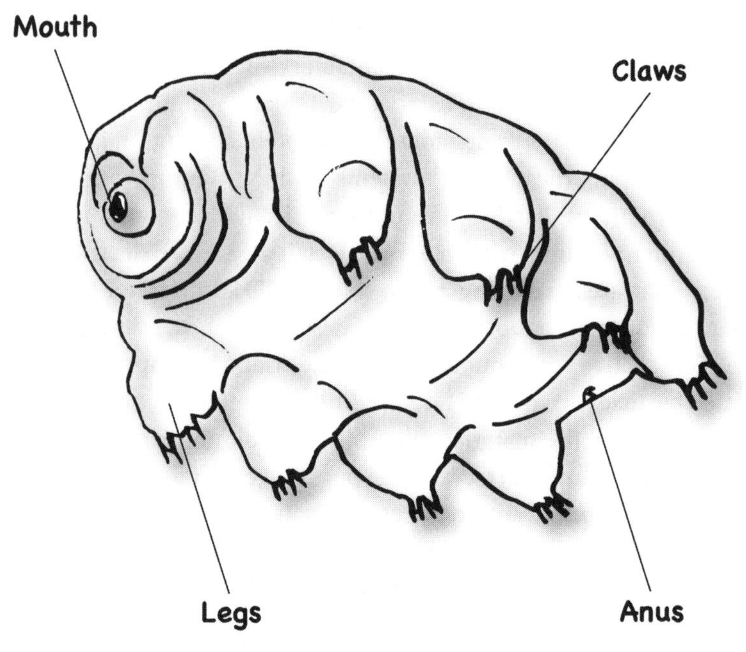

The cuddly(?) tardigrade or 'water bear'.

zero. Some can withstand pressures around six times greater than those found in the deep-sea trenches, such as the Mariana, and radiation exposure at hundreds of times higher than a human can.

Unlike bacteria that live in hot springs, which are considered *extremophiles* because of their adaptations to reside in these (to us) extreme conditions, tardigrades merely endure them, so they are not regarded as extremophilic organisms. They are one of only a handful of known creatures that can suspend their metabolism, a process known as *cryptobiosis*. If only I could suspend my metabolism. I'm guaranteed to be hungry again an hour after eating; some tardigrade species, however, can go without food for decades in a state of dormancy. Imagine going to sleep for 20 years without a snack! When in a state of cryptobiosis, tardigrades can lose up to 99% of their water content and still rehydrate and live normally again[9] – when the conditions become favourable.

In 2019, an Israeli space mission called 'Beresheet' attempted to land on the Moon.[10] The lunar lander carried a time capsule with over 30 million pages of data, including memoirs of Holocaust survivors, children's drawings and the English-language Wikipedia, along with the Israeli Declaration of Independence. Its scientific payload included a magnetometer and equipment to measure the precise distance between the Earth and the Moon. However, genetic samples and a group of our tardigrade friends were added to the capsule at the last minute. On 11 April 2019, the lunar lander crashed at incredible velocity on the Moon's surface. Some scientists at the time suggested that the tardigrades might have survived the crash and lived on the Moon for a while. This sparked a debate about the lack of international regulations around interplanetary contamination with earthly creatures.

So, could the tardigrades survive such an impact on the Moon? Inspired by this very question, Alejandra Traspas and her PhD supervisor, Dr Mark Burchell, ran some innovative tests to find out.[11] They fed 20 tardigrades with moss and water and put them into their dormant state by freezing them for 48 hours. They then put them into a hollow nylon bullet and loaded them into a special experimental gun. They shot the tardigrade-harbouring bullets into a sandpit. They discovered that the tardigrades miraculously survived the impacts of up to 900 metres per second, with shock pressure of 1.14 gigapascals. This is over 2,000 miles per hour. It is unlikely a human would survive being hit by a car at a mere 50 miles per hour. This highlights the extraordinary abilities of the tardigrade to absorb incredible impacts. But was this enough to protect the tardigrades from the impact of the fateful lunar crash? The researchers think not. Above 900 metres per second, the tardigrades collapsed. The impact of the lander during the crash was much higher than this. This study demonstrates that panspermia is difficult to enact. Yet crumbs breaking off meteoroids upon entry could experience far lower shock pressures. Moreover, some microbes can survive impacts of up to 5,000 metres per second, so it's by no means impossible.

With all of the tardigrades' extraordinary characteristics, perhaps it's no surprise they are the first known animal to survive exposure to outer space. In 2007, a group of scientists blasted dehydrated tardigrades off to space on the Foton-M3 mission. They exposed the tardigrades to the hostile conditions of outer space for ten days. Back on our blue dot, nearly 70% of these tardigrades were reanimated within 30 minutes and still produced viable embryos. This incredible resilience has inspired scientists to explore controversial applications, such as enhancing human genes using genome-editing tools. The aim is to increase human resilience to the harsher conditions of other planets, such as Mars – no, it's not a sci-fi novel – and scientists have already experimented by splicing tardigrade genes into human cells in the laboratory.[12] The result? The engineered human–tardigrade hybrid cells exhibited greater resistance to high radiation levels than the normal human cells. Whether or not this is safe enough yet to be replicated in living human bodies is not possible to say. But scientists think that genetic engineering could improve our capacity to colonise other space rocks one day. Personally, I think we should focus our efforts on strengthening our capacity to protect and respect our wondrous and beautiful Planet Earth.

CHAPTER 15

You Are What Your Microbes Eat

'To ferment your own food is to lodge a small but eloquent protest – on behalf of the senses and the microbes – against the homogenization of flavours and food experiences now rolling like a great, undifferentiated lawn across the globe.'
—Michael Pollan

Vegans and dairy consumers alike, would you drink cow's milk without the cow? Israeli-based company Imagindairy recently announced it had raised US$13m in seed funding.[1] Investors had taken a punt on transforming the milk industry. Imagindairy plans to make traditional milk products from microbes instead of cows. The microbe-cow milk produced would be identical to cow's milk in taste and molecular structure. The cow, along with her greenhouse gases, would be replaced by microbes genetically reprogrammed to produce milk proteins – precisely the same proteins as a cow would produce. In the lab, the company's scientists developed a method to insert molecular instructions for the cow's milk proteins, namely whey and casein, into fungi and plant-associated microbes. Then, they threw some plant-based nutrients and water into the mix, and voila, cow's milk without the cow. The company says it 'feeds microbes instead of cows' and that the microbes are up to 20 times more efficient than cows at converting the food into dairy products. The broader process, *precision fermentation*, has been used for 40 years to produce complex organic molecules like the rennet enzymes used to make cheese. Now it's being used to create animal-free milk proteins.

Because the process skips the need for cows, it could have beneficial environmental impacts by reducing greenhouse gases, the need for nasty antibiotics, and habitat destruction. Rearing dairy cows is responsible for 2.1 gigatonnes of carbon dioxide per year and a huge amount of the more potent methane gas.[2] Cutting methane is a key priority in the fight against climate change. In recent years, the alternative milk industry has seen a boom in sales, with a considerable variety of plant-based milk hitting the market, including soy, pea, oat, rice and almond milk. Imagindairy said a recent advancement allowed them to isolate the cow's milk protein from the microbes, which streamlined production and made it easier to scale up. The microbe-cow milk still requires water and land. So, how does it stack up against the traditional dairy footprint? Well, according to Imagindairy founder Eyal Afergan, the milk only needs 10% of the water and 1% of the land required by traditional dairy cow milk. If the public gets behind this and the company produces commercially sustainable levels, it could be a game-changer. The term 'flexitarian' has become a popular label for people who endeavour to consume less meat and dairy. In a recent European survey, nearly a quarter of participants said they were 'very likely' to eat 'fake' cheese if it was identical to conventional cheese.[3] So, it seems there is a market. The alternative cheeses for sale have greatly improved. To begin with, they tasted like anything between yeasty rubber and malty cardboard. The protein in alternative cheeses is currently very low, but this looks likely to change with the help of our invisible friends.

Given the biodiversity and climate change crises, I think strategies like this could make a big difference. Nearly 80% of land cleared for agriculture is for livestock farming.[4]

There is a fundamental principle in ecology called the '10% rule'.[5] It states that when energy is transferred from one trophic level to the next, only approximately 10% of the energy will be passed on. A trophic level refers to the position of an organism in the food chain – often presented as a trophic pyramid. The pyramid has between four and six levels depending on the ecosystem and the perspective of the ecologist who describes it. Here I'll use a five-level pyramid (see below). At the bottom of the pyramid, we have *primary producers*. These are typically classed as the photosynthesisers – the plants and phytoplankton. The next level up has the *primary consumers*. These are the herbivores that feed on the plants. Then we have *secondary consumers* on level three, such as the carnivores that eat the herbivores. Then we have the *tertiary consumers*, which consume the carnivores, and finally, the *apex predators* – the top of the food chain. Typically, microbes are lumped into a separate category, the *decomposers*. I would argue that many microbes are primary producers that feed the plants – such as nitrogen-fixing and nitrifying bacteria. But anyhow, this is how the trophic pyramid is traditionally presented. Regarding the 10% rule, when the herbivores eat the plants, the transferred energy goes from 100% to 10%, and the rest is lost as heat. As we step up from the bottom of the pyramid and the primary consumers eat the herbivores, the remaining energy goes from 10% to 1%. Then from 1% to 0.1%, until you reach the apex predators, where only 0.01% of the original energy is transferred!

This is important in the microbe-cow milk story because the land and energy required to feed the cows are extremely inefficient, as only 10% of the energy from the feed crop is passed on to the cows. The microbe-cow milk reduces this problem. I feel the need to reiterate that not all livestock farming needs to be detrimental to our planet. The regenerative agricultural methods I discussed in an earlier chapter are a testament to this, whereby grazing animals can help restore the land. Moreover, many communities rearing small-scale cattle farms contribute far less to environmental degradation than larger commercial operations.

It is interesting to see how microbes are segregated and lumped into the 'decomposer' category in these typical trophic pyramids. There is an argument that microbes should be firmly recognised as critical players at the primary producer level. Or even at a level before the primary producers – as nutrient providers, communication facilitators and protectors, not simply as decomposers. Perhaps more radically, there is also an argument to re-envision the pyramid completely in light of holobiont research. After all,

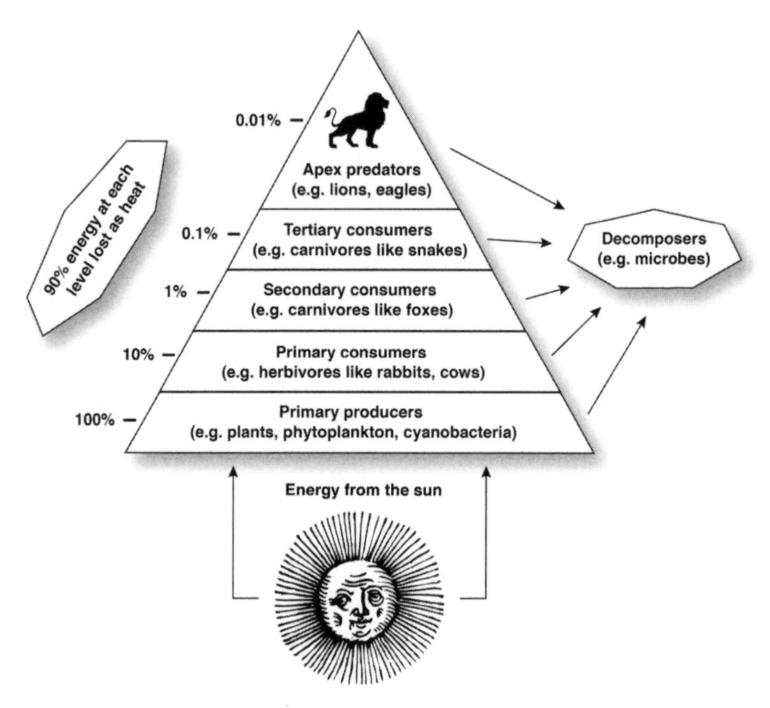

The trophic pyramid.

from a holobiont perspective, where microbes and hosts together act as a whole, trophic levels become fuzzy, and microbes become central to each level in the pyramid.

Even if the microbe-cow milk and cheese are not your bag, well, regular cheese also requires microbes. The identity of most cheeses is intimately dependent on the microbial communities that make the cheese. Thermophilic microbes such as *Lactobacillus helveticus* and *Streptococcus thermophiles* are vital in producing Italian and Swiss cheeses. At the same time, microbes that ferment lactose at lower temperatures, such as *Leuconostoc mesenteroides* and *Lactococcus lactis*, are essential in the production of Cheddar and Gouda. Cheese producers introduce microbes throughout the various stages of cheese production, from the initial milk stage to the fermentation, ripening and ageing.

The art of cheese making has been practised for thousands of years – probably more than 9,000. Cheese is a living conglomerate,

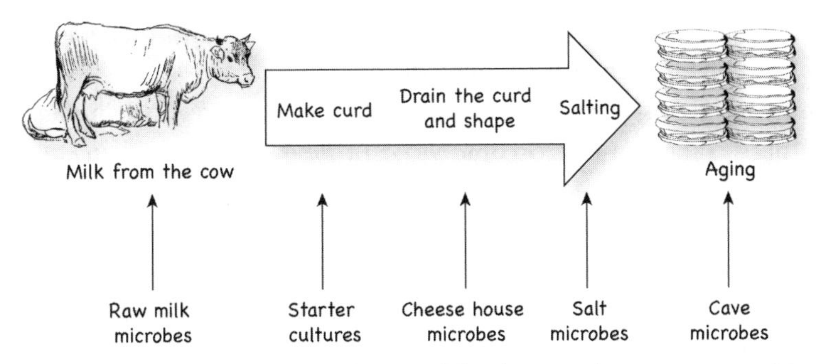

Different microbes colonise at each stage of the cheese-making process, resulting in unique flavours, smells and textures.[6]

brimful of microbes, and the ripening process and wonderfully rich flavours result from the natural succession of microbes that colonise the cheese. The microbes contribute to the final flavour, smell, texture and even colour of the cheese. It is incredible to think about the foods that trigger waves of sensory delight the moment we put them into our mouths, and how microbes are indirectly and directly responsible for this. Still, even before this moment, we *crave* the foods, which triggers a tsunami of dopamine in our brains – the 'feel-good' neurotransmitter – and again, microbes play a role in this craving. Research has shown that the brain releases dopamine upon receiving a sensory reward. But it also releases dopamine in anticipation of the reward. When you associate an activity with pleasure, the anticipation triggers the dopamine response. Robert Sapolsky tells the story of monkeys who were given a reward once they did some work after a signal was activated.[7] Most people assumed the dopamine levels in the monkeys would only increase once they were given the reward. However, this was not the case. The dopamine levels in the monkeys shot up when the signal was activated and maintained at a high level whilst doing the work, but then dive-bombed once the reward was given. This shows us that dopamine is reactive to the *pursuit* of happiness, not just the happiness itself! So, microbes can trigger the release of dopamine in the gut and indirectly trigger dopamine secretions in the brain by instigating the craving for delicious microbe-derived foods such as cheese.

Since ancient times, humans have used microbes to make some of our most wonderful foods and drinks. We owe microbes not only for giving us cheese but for bread, yoghurt and wine, amongst others. Take the microscopic yeast (a type of fungus) *Saccharomyces cerevisiae*, for example. Humans have exploited this fungus in baking and brewing for millennia. Surely this microbe must be at the top of many people's lists of favourites. If you don't have a favourite microbe yet, may I suggest you consider this one? *S. cerevisiae* has been isolated from soil, trees and ripe fruits, and it is used to ferment wine, beer, cider, sake, bread and cakes.

Many scientists date the discovery of yeast's role in fermentation to the mid–late nineteenth century. Yet, people have used yeast in baking and brewing for thousands of years. There are records of Ancient Egyptians using yeast to produce rising bread around 4,500 years ago.[8] And, until recently, many accepted the artificially constructed notion that the Spanish brought grapes over to the Americas and introduced the native peoples to alcohol. In truth, American Indigenous Peoples had been fermenting corn to produce a type of beer for centuries before the Spanish rocked up with their Tempranillo.[9] Historically, there has been a tendency in the West to eradicate traditional knowledge and downplay the abilities and richness of Indigenous cultures. For example, as Bruce Pascoe revealed in his fantastic book *Dark Emu*, Indigenous Australians have been making a form of bread from kangaroo grass seeds for thousands of years – they were possibly the inventors of bread.[10] Their highly organised cultivation practices also question the long-held notion that the dawn of agriculture occurred in the Middle East. I would not be surprised to find out that Indigenous Peoples produced a diverse bounty of yeast-aided products across the planet, and way before European ale and bread. However, in Europe in the mid-1800s, Louis Pasteur and colleagues advanced the fermentation process by culturing purer yeast strains. Industry leaders introduced huge growing vats to Great Britain, and growing yeast became an enormous industrial endeavour.

The name *Saccharomyces cerevisiae* comes from the Latin for sugar (*saccharo*), fungus (*myces*) and beer (*cervisia*). It is present in the environment, but it is not airborne; thus, it requires a vector to carry it from one environment to another to secure its genetic diversity, and ultimately its survival.

Cue the social wasps. Several studies have shown that social wasps are likely to be key players in the dispersal of *S. cerevisiae*.[11] It is well established that grapes undergo fermentation naturally, but the yeast is seldom present on unripe grapes. The social wasps are thought to transfer the yeast onto the grapes, thereby facilitating fermentation. One study found that the yeast is a constant member of the wasp gut microbiome.[12] The wasps feed on the grapes when they are nice and ripe. When the wasps pierce the grapes, the yeast transfers from their guts to the grape, instigating the natural fermentation process. The researchers also discovered that hibernating females harboured the yeast cells in their guts and passed them on to their progeny the following spring by regurgitating food. This is incredible; insects, particularly social wasps, likely play a vital role in the winemaking process. We have the wasps (and the yeast!) to thank for our wine's natural fermentation and complexity. Incidentally, it was recently found that *S. cerevisiae* shapes the gut microbiome in social wasps and enhances their adaptive immunity to pathogens such as *E. coli*.[13] These bacteria can cause nasty infections in the wasps. A potentially important selective mutual advantage is likely at play here – the wasp depends on the yeast microbe for immune priming, and the microbe depends on the wasp for dispersal… and we depend on both for our wine.

Some researchers have argued that notable changes in the early hominin genome occurred millions of years ago. A gene that encodes alcohol dehydrogenase, which breaks down alcohol in the liver, evolved around ten million years ago when early apes left the treetops to walk on the ground.[14] Because of this behaviour change, the apes would encounter more ethanol-rich fruits that would fall from the trees and ferment below. Researchers recently pointed out that modern-day great apes such as chimpanzees enjoy a nice tipple of alcohol and even make tools to access the alcohol from the fruits. Other researchers argue that our early hominin ancestors likely fermented fruits using yeasts like *S. cerevisiae* a million years ago.

If you're a chocoholic like me, you have our invisible friends to thank (or should that be to blame?). An ordered natural succession of microbes is responsible for the fermentation of cocoa beans, and without it, we would not have delicious chocolate. Before the sensory pleasure of eating chocolate fills our minds, the production of chocolate involves a long chain of events that most of us do not witness or likely consider. The cacao tree, *Theobroma cacao*, is about 6 metres tall when cultivated – though it may reach over 18 metres in the wild. The cacao fruit is a rugby-ball-shaped pod containing the beans. After the pods are harvested, highly skilled, machete-wielding workers split them. Then, the workers gently scoop up to 50 beans out of a single pod, and one worker can harvest enough pods to produce over 50 kg of chocolate per day. Then comes the all-important fermentation process, where microbes are the stars of the show.

Fermentation dramatically reduces astringency and transforms the flavour into what we now associate with chocolate. The process takes several days to complete. Incidentally, the pulp that drains off the beans during the fermentation process, which is given the delightful name of *cocoa sweatings*, can be partnered up with wine yeast to make cocoa wine.

The first stage of cocoa fermentation is the *anaerobic phase*. This involves a 48-hour period during which microscopic yeasts help transform the sugar of the cacao pulp into alcohol and carbon dioxide. It is thought that fruit flies are responsible for transferring microbes to the beans. Flies of various forms are also vital for pollinating the cacao trees. You will notice a pattern here – microbes and insects are the underappreciated pillars of society.

The second fermentation stage is the *aerobic phase*. During this time, the cacao beans are turned and aerated. This allows *Acetobacteria*, which are oxygen-loving species, to transform the alcohol left by the yeasts into acetic acid. Temperatures begin to rise, and the heat and acetic acid kill the cacao bean bud. At this stage, a swathe of changes occurs through chemical reactions and enzyme secretions in the cacao bean. These microbe-induced processes eventually develop the chocolate flavours we so love. Banana leaves are often placed on top of the cacao beans to help

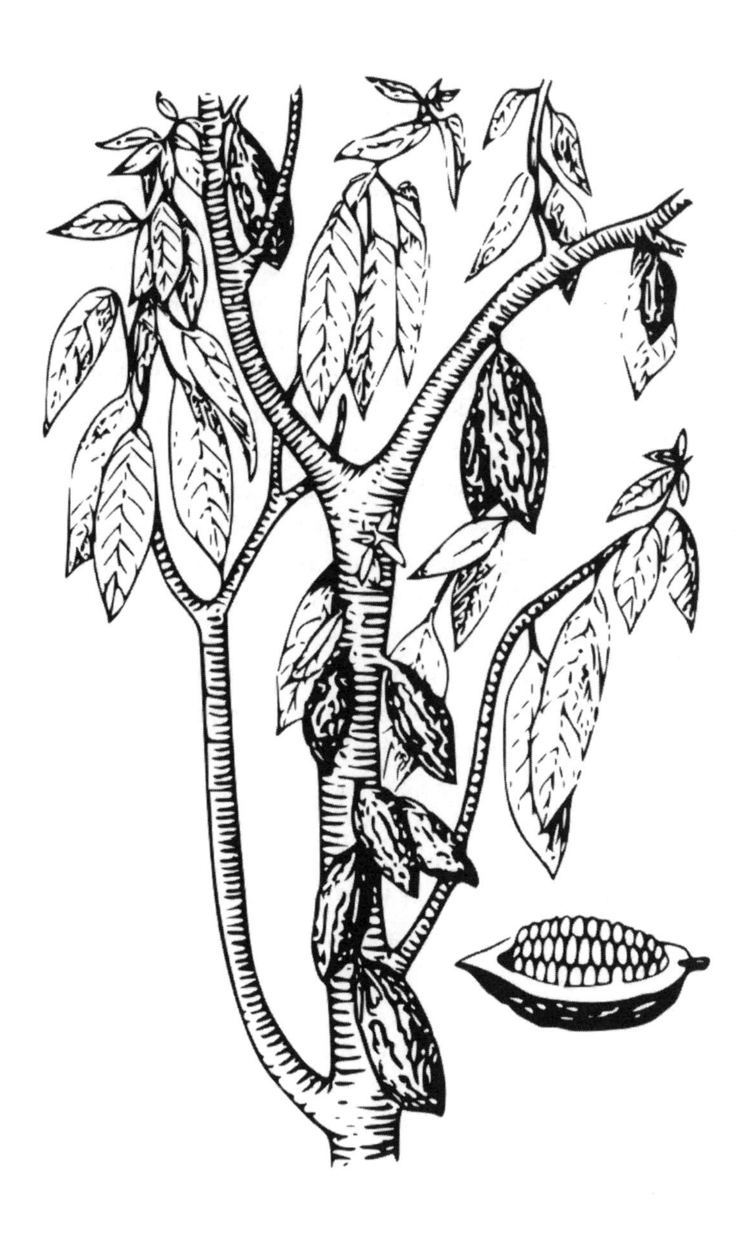

Cacao tree and a pod full of cacao beans.

facilitate the fermentation. The banana leaves conserve the heat whilst providing additional yeasts and bacteria, strengthening the fermentation process. A suite of actions follows to complete the chocolate-making cycle, including drying and roasting.

Coffee. Ah, coffee. That most wondrous and ubiquitous of legal drugs. Legend has it that a goat herder called Kaldi first discovered the potential of coffee beans in the ancient coffee forests of the Ethiopian plateau.[15] His goats would graze on the beans from the trees, and they became so stimulated and full of energy that they seldom slept at night. Kaldi shared his findings with the local monks, who brewed the beans, and to their delight, it kept them awake during long periods of prayer. Word spread like wildfire, and coffee cultivation began on the Arabian Peninsula in the fifteenth century. By the mid-seventeenth century, coffee had replaced the traditional breakfast beverages of the time in Europe – beer and wine! Like in chocolate, beer and wine, microbes such as yeasts and bacteria are essential in coffee bean fermentation. We know lactic acid bacteria play an important role, but we still don't fully understand the complexity of the different roles microbes play. Microbes produce some of the rich and complex scents that come from the coffee bean's volatile compounds. Lactic acid bacteria contribute to the fruity notes and may help protect the coffee beans from undesirable microbes that can lead to unpleasant flavours. Many microbes contribute to the fermentation process, including bacilli, acetic acid bacteria, enterobacteria, yeasts and filamentous fungi. Research is underway to find out exactly how they contribute.

The next time you're having breakfast and sipping your coffee, take a moment to appreciate the complexity of the coffee-making process and the different roles microbes play in the relationship between your desires, your behaviour and your chosen drink. It's all connected.

The next story is about kimchi and kombucha. Kimchi is a Korean dietary staple, an emblem of fermented foods, a jewellery box of diverse microbes. Koreans eat two million tonnes of kimchi each year – that's 40 kg per person. It is known that specific microbes play a crucial role in kimchi's structure and taste, but only recently has an exploration of the microbial community of kimchi taken place. Using four key ingredients of kimchi – cabbage, garlic, ginger and red pepper – researchers selectively sterilised each ingredient to see which one contained the *Lactobacilli* bacteria needed for fermentation. They found that cabbage and garlic both contributed these bacteria, whereas red pepper and ginger did not. It is thought that red pepper and ginger can control unwanted microbes and boost the growth of the friendly lactic acid bacteria. None of the ingredients alone built the complex bacterial community and the best kimchi; every one of them is required to work together, to collaborate, cooperate and produce the complex and dynamic community necessary for the kimchi to flourish as a mini ecosystem.

During harsher times in the Korean nations, fermented foods such as kimchi allowed people to preserve and store their food for longer. Kimchi is also packed full of nutrients and probiotics. Some evidence suggests that regularly eating kimchi could help prevent diseases by reducing inflammation and boosting your immune system – thanks in part to the microbes and their by-products.[16] So, microbes are vital in kimchi production, and consuming microbe-laden food could also benefit the human microbial ecosystem.

Kombucha is a fermented tea beverage with an effervescent taste. It's composed of a complex microbial ecosystem characterised by cooperation and competition. The origins of kombucha are not certain, but some reports suggest it originated in Northeast China around 2,000 years ago and was traditionally prized for its healing properties. Records indicate the drink was produced in Russia in the early 1800s, a sign that trade routes had expanded across Asia and Europe, allowing the drink to percolate across cultures and continents.[17]

A rich community of bacteria and yeast initiates the fermentation of black or green tea leaves to make kombucha. After a short while, a biofilm is produced, covering the liquid. This occurs through several phases during which microbes in the solution cooperate to metabolise the sugars. They also mount a group-based biofilm response to protect against invading microbes and to create storage for resources. Ethanol and acids are also produced during fermentation, which help protect the community against microbial competitors from the environment. The kombucha fermentation process is characterised by a diverse orchestra of microbial cooperators, working mutualistically to sustain a tiny yet bustling ecosystem. Humans could learn from the fermentation process: social change has parallels with this concept of mass cooperation for shared benefits.

Much like kombucha, sourdough is an ancient food product packed with microbes from different kingdoms of life. Sourdough requires a starter culture of flour, water and microbes. Once a starter is made, it can be incredibly stable. Some sourdough starters have been passed down from generation to generation. In 1889, a year before Wyoming became an official US state, a group of shepherds in a village called Kaycee began growing a sourdough starter.[18] The starter was passed down over many generations until a man who acquired it from a student at the University of Wyoming gave it to his daughter. Lucille Dumbrill was her name. In 2001, Lucille was featured in the local newspaper, as she had kept the culture in her refrigerator for decades and was still using it for baking fresh sourdough bread. By this time, the starter was 112 years old, and it's still believed to be around and active today. Lucille, now well into her nineties, plans to pass it down to her daughter, who will hopefully help keep the sourdough alive for generations to come. A San Francisco bakery claims it has been using the same sourdough starter since it opened in 1849, making it over 170 years old![19]

Once flour and water are mixed in a bowl, almost immediately, the paste is colonised by bacteria and yeast from the environment, and the microbes begin feeding on the sugars. Many microbes can grow in this recently formed paste, including unwanted bacteria which can spoil it. However, before long, the conditions start to change. Lactic acid bacteria are amongst the early colonisers, and they begin to acidify the starter. A few days in, and these bacteria have increased the acidity of the starter so much that many of the early opportunists – including the spoilers – can no longer survive, and only the lactic acid bacteria and a couple of acid-loving yeasts remain. This acid, along with the acetic acid produced by other bacteria, gives the sourdough its characteristic tangy flavour. A by-product of the yeast is carbon dioxide, which creates airy pockets and the rise in the bread.

A couple of weeks in, the starter has entered a more stable state whereby *Lactobacilli* and the yeasts grow vigorously. Even after this increased stability, microbes in the sourdough community still come and go. Lucille's starter community will likely be slightly different from that of the starter way back in nineteenth-century Kaycee.

During the recent COVID-19 lockdowns, sourdough baking became a global trend. It's creative and yummy, but it also brings people together, sharing starter cultures, passing invisible friends on between baker friends. We live in times when COVID-19 has understandably led many people to fear and loathe microbes. Yet sourdough is a ray of light, showing us that microbes also do wonderful things for society.

The term 'you are what your microbes eat' has multiple meanings. From one perspective, all the foods I've discussed in this chapter depend on the microbial by-products created when the microbes 'eat' the ingredients. On the other hand, your entire diet feeds the microbes in your gut, which in turn release myriad chemicals that sustain your health. I've already mentioned how certain gut microbes can produce health-promoting by-products such as the short-chain fatty acid butyrate. They do this by digesting the high-fibre foods we eat, such as fruits and vegetables. So, what more do we know about how our diet affects our microbes, which also affects our health and behaviour?

The concentration of dietary fibre is much lower in processed foods, such as biscuits, fast food and confectionery, than in fresh, raw ingredients like fruits, nuts and vegetables. Gut microbes devour fibres in the gastrointestinal tract, and some digest carbohydrates, such as starch and non-starch polysaccharides. These cannot be directly metabolised by the human host alone. These complex nutrients act as a vital energy source for microbial growth. Fibres and carbohydrates are also known as 'prebiotics'. Prebiotics induce essential changes in the composition of the gut microbiome, which influence our health and behaviour. It is widely acknowledged that the consumption of the modern Western diet, which is often heaving with processed foods, can result in the loss of important microbes. Researchers have found that individuals whose diet is lower in saturated fat and higher in fibre tend to have fewer pathogenic bacteria and more beneficial bacteria.[20]

The Mediterranean diet is considered by many to be an emblem of a healthy lifestyle. It's predominantly based on a diverse mixture of fruits, vegetables, grains and polyunsaturated fats. This diet is known to have significant anti-inflammatory properties, which may be related to interactions with the gut microbiome. Studies have also shown that children with vegetarian diets rich in plant-based polysaccharides and fibres have significantly more bacteria associated with protection against inflammatory diseases.[21] The gut microbiome composition differs among individuals living in different geographic regions and between individuals of different cultures. This suggests that cultural practices influence the gut microbiome, probably through different dietary choices and other lifestyle phenomena.

Those all-important short-chain fatty acids are released because of fermentation in the gut – perhaps I should have called this chapter 'Fermentation'! These acids lower the pH of the colon, which in turn has a major influence on the type of microbes that reside in the gut. This can be protective; for example, a lower pH can limit the growth of some pathogenic bacteria such as *Clostridioides difficile*, which can cause nasty symptoms like diarrhoea and sickness. Foods rich in the necessary fibres or prebiotics include bananas, leeks, garlic, asparagus, leafy greens, seaweed, onions and beans, along with whole grains and seeds like oats and quinoa.

Fermented foods such as kimchi and kombucha contain live microbial communities and undergo remarkable alchemy to benefit our human ecosystem. This chapter only mentions a few of the fermented foods out there. Many others exist, such as kefir (fermented milk) and tempeh (fermented soya beans). It is thought that fermented foods can help restore balance in the gut ecosystem and improve the availability of nutrients. There's even some evidence to suggest probiotic bacteria found in fermented foods may improve symptoms of anxiety and stress. Including fermented foods in your diet in the long term may enhance your life expectancy when part of a broader healthy lifestyle. A recent study revealed that South Korean women born in 2030 would have a 57% chance of exceeding the age of 90 and a 97% chance of living over the age of 86.[22] Some experts believe that the Korean diet, which is rich in kimchi, and thus both pre- and probiotics, could make an essential contribution to their long lifespan.

Short-term, fleeting dietary choices will probably fail to impact our health, particularly in positive ways. Further investigations to identify the effects of long-term dietary patterns on the gut microbiome are greatly needed. Yet it is likely that committing to a long-term change in your diet will be required to reap long-term benefits. Moreover, many other environmental and lifestyle choices will affect the human microbial ecosystem. If these are not recognised and accounted for, studies assessing the impact of food on the gut microbiome may result in confusing outcomes and delusive generalisations.

Combining a long-term, healthy and diverse diet with spending more time in biodiverse environments is the way to show respect to your microbes. They'll reciprocate by making sure your internal microscopic ecosystem flourishes. There are many parallels in the macro world; an old-growth forest takes a long time to establish its diverse communities and complex interactions that ultimately sustain a flourishing metropolis of life. In an ideal world, we would establish a similar human ecosystem from *early life*. Yet if this has passed you by, there's always the principle of ecological restoration to embrace. Reduce the harmful exposures (antibiotics, processed foods, other noxious chemicals), immerse yourself in the diverse life-forms of natural environments, reconnect with the wider biotic community – psychologically, emotionally and biologically – and gradually create the conditions for life and health to flourish. Rewild your body. Rewild your mind. Love your microbes, and they will return the gift.

I will now finish this chapter by sharing one of my favourite microbe-loving fermented food recipes that, if you wish, you can try at home.

Spicy fermented vegetables

I've discovered many fermented vegetable recipes online and have created my own, inspired by these.[23] Here's one of my favourites – it's a hot one! It combines a few different vegetables with some chillies, spices and peppercorns – perfect for those who prefer something nice and spicy. If you're not a fan of the spice, you can simply leave these out of the mixture.

Ingredients

½ teaspoon black peppercorns
½ teaspoon coriander seeds
½ teaspoon caraway seeds
2 teaspoons salt
2 red chillies
6 slices ginger root
500 ml water kefir to submerge the vegetables
3 medium-sized carrots
1 stalk celery cut into 7 cm length
2 bok choy or savoy cabbage leaves and stems

Instructions

1. Clean a 1 litre glass Mason jar, fitting lid (plastic preferred due to the acid produced during the fermentation that can corrode metal), and other utensils with soapy water and rinse with tap water. No sterilisation is needed.
2. Pack the Mason jar with vegetables as tight as you can.
3. Add the remaining ingredients.
4. Add more water kefir to ensure the vegetables are immersed in liquid.
5. Cover the Mason jar tightly with a lid. Keep the jar at room temperature and allow the vegetables to ferment.

6. The bok choy leaves can be ready in 2 days, while the carrots and celery can be ready in 5 days.

Enjoy!

You can continue to marvel at the power of microbes whilst enjoying your fermented food for years; simply add more of the raw vegetables and liquid continuously into the Mason jar to keep the bacteria alive and kicking. It is thought that the longer the probiotic brine (the liquid in the jar) is used, the stronger and more complex the flavours become.

CHAPTER 16

Nature Connectedness

'Humankind has not woven the web of life. We are but one thread within it. Whatever we do to the web, we do to ourselves. All things are bound together. All things connect.'
—Chief Seattle

It was around 2006. I sat on a balcony, almost cradled by overhanging oak tree branches. A book rested in my hand, and my mind was treated to the pages of *The Diversity of Life*, eloquently penned by the late naturalist E.O. Wilson (1929–2021) and first published in 1992.[1] I remember feeling a strong emotional connection with the natural world, and this majestic concoction of ecological and evolutionary insight was balm to my soul. I had, it seemed at the time, sizeable life decisions to make. There's no escaping the mawkish sentimentality of this, I admit, but it felt as if nature and literature came together in one moment to lay down the first bricks in the road that would for ever change my life. With a still-developing prefrontal cortex and perhaps a hangover from adolescent naivety, there is no doubt I was an impressionable soul back then. However, it is still easy for me to see the profound influence and inspiration that nature bestows.

> Somewhere close I knew spear-nosed bats flew through the tree crowns in search of fruit, palm vipers coiled in ambush in the roots of orchids, jaguars walked the river's edge; around them eight hundred species of trees stood, more than are native to all of North America; and a

> thousand species of butterflies, 6 percent of the entire world fauna, waited for the dawn. About the orchids of that place, we knew very little. About flies and beetles almost nothing, fungi nothing, most kinds of organisms nothing. Five thousand kinds of bacteria might be found in a pinch of soil, and about them we knew absolutely nothing. This was wilderness in the sixteenth-century sense, as it must have formed in the minds of the Portuguese explorers, its interior still largely unexplored and filled with strange, myth-engendering plants and animals. From such a place the pious naturalist would send long respectful letters to royal patrons about the wonders of the new world as testament to the glory of God.[2]

Spending time in wild places has always been a passion of mine. Still, in my twenties, I quite suddenly switched to what I can only describe as *hyperdrive mode* – perhaps influenced by several consciousness-expanding experiences. One of the most profound of these admittedly arrived at the hands of the natural psychedelic compound psilocybin (consuming fresh 'magic mushrooms' was legal in the UK at the time). Yet, although it may seem like I am glazing over the subject, I can say with certainty that even simple, mindful connections with nature have led to several powerful, perspective-shifting experiences. As a result of these, I developed a new-found determination to live life to the fullest. For me, this included learning as much as possible about the vast array of biological diversity and the natural processes that have fascinated me since I was a child. In learning more about the natural world – including its diverse microbial inhabitants – I also became fascinated by the bounty of edible wild plants that grow in urban and rural areas. The same ecological theatre supporting those euphonious birdsongs and alluring floristic blossoms is also an outdoor greengrocer, brimming with weird and wonderful nutrient-rich ingredients! We inadvertently stroll past many of these ingredients each day, sometimes on the way to the shops; indeed, sometimes on the way to the shops to purchase the same items we just walked past, such as blackberries, mint, hazelnuts, rocket and herbal teas. Foraging brought me closer to the land. It also brought me closer to the environmental microbiome.

Each time I pick a blackberry, I receive a suite of microbes from the berry and all the butterflies, hoverflies, beetles and bees that visited the berry before me. As I climb the limestone cliffs to pick some wild oregano for my pizza, I receive a suite of microbes from the rocks, the mosses, the knapweed and all the creatures that visited them before me. I fill my lungs with naturally diverse bacteria, archaea, fungi, viruses, algae and protozoa that float off the plants and the soil. It is possible that this exchange of invisible friends boosts my immune system; perhaps in the long term, the spore-forming butyrate producers help reduce inflammation and contribute to cell signalling deep inside the caverns of my body, and perhaps this experience reconnects me with those mighty microbial 'old friends'. Whatever happens in these moments, I know my personal ecosystem changes. I know that if I do this regularly, my mind and body are strengthened, making me more resilient to modern maladies. I know that evolutionarily, this is where I am supposed to be and that these correspondences with nature inspire me to protect it.

There is a concept known as *plant blindness*, or *plant awareness disparity*, defined as the inability to see plants in one's environment.[3] This may seem strange, but it is thought that many of us simply do not notice the plants around us – possibly due to the increasing demands for our attention at the hand of modern living. For example, when we repeatedly check our smartphones, ruminate about deadlines and performances or crave new purchases, we may also lose touch with our surroundings. Several authors have written in depth about the growing disconnect between humans and the rest of the natural world. Ironically but also aptly, a recent survey carried out on a smartphone app found that 50% of children were unable to identify a common stinging nettle.[4] This is a ubiquitous and edible perennial herb found across the Northern Hemisphere.

It's also covered in tiny hairs that, when brushed against, inject a potent concoction of histamine, acetylcholine and formic acid to produce a stinging pain – hence the name. So, in theory, it should be extra memorable.

However, there is hope. I have seen a resurgence in people becoming interested in nature – or as Professor Miles Richardson would say, 'the rest of nature'[5] – whether it's to get involved in health and well-being activities like *shinrin-yoku* (forest bathing) or going on a local foraging adventure to create new and exciting recipes. This is one of the reasons for writing this chapter – to help keep the wheels turning.

Importantly, as I mentioned in the Introduction, we now have a profound opportunity to redefine our relationship with nature not only culturally, socially, psychologically and emotionally, but *microbiologically* – through knowledge of the unseen, nested layers of nature inside and around us. Immersing ourselves in the microbial biodiversity we co-evolved with not only benefits our holobionts (our walking ecosystems), but as a mindful experience, it can expand our previously parochial understanding of the world.

The contents of this book are a testament to the knowledge we now have of this unseen cosmos. We have microscopes, DNA sequencing, visualisations and rich stories to feed our perceptions. Our limited visual acuity can no longer restrict our ability to appreciate the invisible kingdoms of life – only the willingness of our minds can do this. We have a profound opportunity to learn from the magnificent cooperative interactions of the myriad more-than-human subjects around us. As Robin Wall Kimmerer says, the ethic of reciprocity lies at the heart of the gift economy.[6] We receive innumerable gifts from the unseen constituents of the natural world; now, it's time to give back.

Wander into the natural world and marvel at the sights, smells and sounds. But whilst doing so, also try to remember that inside every tree, in every carpet of moss and every bird that passes, a rich consortium of invisible friends keeps them (and us) alive. Below every step you take, millions of secret life-forms cycle nutrients and shuttle chemicals between the trees. Simply taking a mindful moment to think about this web of interconnectivity can be humbling, and acknowledging its existence and power can be transformative. Ultimately, all the nature you *can* see intimately depends on all the nature you *can't* see. The more we respect the invisible world around us, the more its importance resonates within us. This perspective is valuable because through it, we are more likely to view nature – including ourselves – as a complex web of interrelated subjects and lock away the dominionistic mindsets that have cursed some cultures in the past and present. Understanding that all things are connected, in part through our invisible friends, also paves the way to create the conditions that allow the health of all organisms to flourish. The more you think about the unseen, the more the veil is lifted away.

Continuing with the theme of connectedness, *nature connectedness* (also known as *nature relatedness*) is an increasingly adopted term in academia, but it's also percolating into the general public's psyche. As mentioned briefly in previous chapters, nature connectedness describes our emotional, cognitive and experiential connection to nature. It's also a recognition that humanity is deeply embedded within nature itself. Nature connectedness is associated with pro-ecological behaviours such as conserving and restoring the natural environment. Researchers have developed validated instruments to measure one's level of nature connectedness. I'll discuss some of the most recent and cutting-edge research on this topic later in the chapter.

Firstly, I'd like to talk about a global threat that is spreading: the fear of microbes, also known as 'germophobia'. This phenomenon could be detrimental to our health and ecosystems by encouraging people to avoid the natural world. It was a remarkable development in human thinking when scientists understood that microbes were responsible for various human illnesses. This has never been more salient than now, as the world battles the COVID-19 pandemic. Microbiology and our knowledge of pathogens have undoubtedly saved countless lives in the last century. However, knowing that some microbes – far fewer than 0.01% – cause human diseases has led many people to fear and loathe all microbes. Decades of advertising campaigns, such as those selling household detergents, have created negative perceptions of microbes as a whole. This has spawned another enemy, but a psychological one – germophobia.

Avoidance of natural dirt and reduced human contact with biodiversity is one consequence of germophobia. It could be contributing to a loss of appreciation for the vital, invisible universe around us.

I once conducted a study on germophobia with colleagues Professor Anna Jorgensen and Dr Ross Cameron.[7] We set out to understand whether there was a relationship between people's engagement with nature and their attitudes towards microbes. We also investigated whether basic 'microbiology literacy', such as correctly identifying different microbial groups, might influence these attitudes. We developed an online questionnaire and received well over 1,000 responses. We found that people who showed positive attitudes towards microbes spent significantly more time in nature per week and spent substantially longer in nature per visit. These results suggest that germophobia-related attitudes may reduce people's desire to spend time outdoors. Or on the flip side, it could mean that spending more time in nature can increase positive attitudes towards microbes. This points to a possible strategy to help challenge the negative consequences of germophobia: spend more time engaging with nature. The benefits of doing so could include enhancing immune function via exposure to environmental microbes, which help regulate our innate immune system (fighting pathogens before they cause infection) and adaptive immune system (stimulating tiny armies of memory cells). There's a range of advantages associated with engaging with nature, making this strategy all the more appealing. For example, it can reduce stress and anxiety while promoting social cohesion and a sense of connection.

We also found that microbiology literacy was associated with positive attitudes towards microbes. This suggests that having a basic understanding of microbes may encourage people to view them in a more positive light. This could be very powerful. For example, teaching children about the different ways in which microbes support individual and planetary health could reduce germophobia in the future by promoting an appreciation for these essential life-forms.

Finally, we found that people who identified viruses as being microbes had a significantly more negative attitude towards microbes in general. This may result from the recent COVID-19 pandemic, as people understandably fear the SARS-CoV-2 virus. However, there is a risk that we may unfairly tarnish all microbes with the same brush. As we have discussed in this book, bacteria, archaea, algae, fungi, protozoa and viruses all have critical roles to play in our ecosystems. A greater emphasis on promoting engagement with nature could help enhance human health and promote more positive, constructive attitudes towards the foundations of those ecosystems – the microbes.

I believe we need to reconnect with the rest of the natural world, not only to rewild our microbiomes but to establish long-term ecological stewardship, thereby protecting the environment and all the microbes that sustain life on Earth. Before I delve too far into nature connectedness research, I feel it is important to acknowledge that many aspects of the nature connectedness concept have been recognised and practised in Indigenous cultures for millennia.

In 1788, a fleet of ships arrived in the Southeast Australian mainland area led by Captain Arthur Phillip. The land was under the custodianship of the First Nation Dharawal Peoples. Many now know the site as Botany Bay, so named because of the huge variety of plants found by the botanist of the ship *Endeavour* and the first president of the Royal Society– Joseph Banks. Banks first stepped onto this enormous landmass during James Cook's famous mission several years beforehand. The 1788 fleet, which eventually settled at nearby Port Jackson, consisted of over a thousand settlers, including 778 convicts. A penal colony was formed, and in the ninety years that followed, it is thought that over 160,000 convicts were transported to Australia.

The penal colonies quickly transitioned to 'free' societies, fuelled by lucrative trade prospects associated with gold-mining and wool production. Records indicate some early white settlers attempted to seek favourable relations with the Indigenous Australians; however, in 1835, the British Colonial Office issued a Proclamation to reinforce the notion that this vast and diverse land had no 'owner' before the British Crown claimed its possession. Widespread conflicts and the spread of European infectious diseases decimated Indigenous Australian populations, and as Bruce Pascoe describes in his book *Dark Emu*, Aboriginal cultures suffered enormously.[8] They suffered to the point where many Aboriginal Peoples were forced to adopt European societal practices or face the consequence of isolation or even death. With this forced cultural transition, rich traditional ecological and tacit knowledge was woefully lost to an unpenned history. Luckily, some remains.

You may now be asking yourself: what exactly does Indigenous Australian culture have to do with nature connectedness? Well, as the historian Bill Gammage suggested in his 2011 book *The Biggest Estate on Earth*, pre-1788 Australian culture was very much infused with a mindset of collective obligation, ensuring the flourishing of all life.[9] This profound cultural evolution fostered an intimate and mutual relationship with nature, a deep connection with the land developed over millennia. Although this holistic concept may seem alien to some, it is also simply a natural and untainted extension of the reality that we are nature and nature is us. Perhaps the West's dominant philosophical and societal paradigms have led to a profound disconnection to the point where we view this archaic holism as novel.

Many Indigenous Peoples across the world share the Indigenous Australians' intimate relationship with the surrounding natural world. This, of course, is a generalisation, but unlike the current typical – or perhaps stereotypical – Western view of nature, many Indigenous groups have historically viewed themselves (and still do) as deeply connected to and very much a part of the natural world. Indeed, we could argue that our growing detachment from this concept, particularly in Western societies, is a symptom of a virulent cultural malady – that is, our perceived dominion over nature.

We can trace this shallow notion back thousands of years. For example, passage 1:28 in the English Standard Version of the Book of Genesis states:

> And God said to them, 'Be fruitful and multiply
> and fill the earth and subdue it, and have

> dominion over the fish of the sea and over the birds of the heavens and over every living thing that moves on the earth.'[10]

However, it is perhaps important to acknowledge that the Catholic Church has since rejected the notion of humankind's dominion over nature, with Pope Francis stating in his encyclical on climate change and inequality, '*Laudato Si*':

> Although it is true that we Christians have at times incorrectly interpreted the Scriptures, nowadays we must forcefully reject the notion that our being created in God's image and given dominion over the earth justifies absolute domination over other creatures.[11]

We could also point to the advent of agriculture, which many scholars suggest started between 10,000 and 20,000 years ago in the Middle East. This progressive control over natural resources could conceivably lead to dominion over nature. However, recent evidence suggests that Indigenous Australians may have cultivated crops and used complex agricultural systems and holistic landscape management techniques for thousands of years. This indicates that agricultural practices per se might not drive the evolution of dominionist behaviours towards nature.

In any case, the image of human dominion over nature can undoubtedly be traced back to the English philosopher and statesman Francis Bacon (1561–1626). During his undeniably stately lifetime, Bacon, with his velvet robe and lace neck ruffle, served as Lord High Chancellor of England and a legal advisor to Queen Elizabeth I. He was an early proponent of empirical scientific methods and an influential figure throughout the scientific revolution. Driven by earlier forms of scientific endeavour and influenced considerably by the Christian religion, Bacon took the image of human dominion over nature as the guiding principle of his new vision of practical knowledge. This vision eventually helped pave the way to modern scientific enquiry. Bacon viewed science as a way to control nature and restore the human superiority lost by Adam and Eve in the fall from the Garden of Eden:[12]

> For man, by the fall, fell at the same time from his state of innocency and from his dominion over creation. Both losses however can even in this life be in some part repaired; the former by religion and faith, the latter by arts and sciences.[13]

In the centuries that followed, talented philosophers and scientists, from René Descartes to Isaac Newton and Charles Darwin to Sigmund Freud, helped seal a depiction of nature subject to mechanical laws that can be exploited, controlled and tamed to provide resources for the benefit of humankind.

But if we are growing increasingly detached from the rest of the natural world, and this self-absorbed view of dominion over nature is at the root, does this matter? And if so, why?

One reason it matters is that the loss of biodiversity (including microbes) – the variety of life on Earth – is now recognised as a global megatrend that is rapidly spiralling out of control, hence the term 'biodiversity crisis'. Current species extinction rates are estimated to be 1,000 times higher than natural background rates, and without transformative solutions, future rates are likely to increase to 10,000 times higher.[14] Importantly, we can attribute this in part to human-driven processes such as unsustainable resource exploitation, pollution and climate change.

To some people, biodiversity holds value in itself and for itself – this is also known as *intrinsic value*. In this philosophical view, all organisms, including microbes, have a value and right to exist regardless of whether they are considered useful or instrumental to humans and their endeavours. The contrasting and largely uncontested view is that biodiversity has *instrumental value*. In this moral camp, nature is seen as only having value in that it offers a means to a desired end for humans. Our vast array of ecosystems undeniably possess myriad valuable facets instrumental to the survival and flourishing of human well-being, such as recreational, medicinal, resource and other types of value that fall under the so-called 'ecosystem services' umbrella (a term which itself is teeming with issues from a dominionistic perspective).

By only viewing the natural world as something to use and control, many humans have forged an unsustainable relationship with nature that, to some, has sparked – metaphorically, but in some cases literally – a blazing fire that is ravaging the livelihoods of species (including ourselves) across the globe. Indeed, every human individual of some 7.8 billion on Planet Earth finds themselves at some level of risk from the current biodiversity and climate crises. Some academics have said, 'There is no human well-being without nature's well-being.'

We may require a transformation of consciousness to solve these issues. Perhaps we need to rethink what it means to be human in relation to the rest of the natural world – a mindful shift towards a new relationship – or perhaps even a rekindling of an ancient relationship with nature could be the critical sustainable solution to these existential risks. In turn, our relationship with the natural world, directly and indirectly, affects our interactions with and perceptions of the microbes that live within it.

One potential approach to re-establishing our connections with the rest of nature is to learn from the rich tacit knowledge and transfer systems of the ancient cultures that still hold this wisdom. However, any engagement with Indigenous Peoples must be done ethically – with respect and remuneration, and in partnership with Indigenous communities.

Another approach that has arisen from an emerging field of research is to investigate and learn precisely how to enhance one's nature connectedness. My first readings of the nature connectedness concept were those of Aldo Leopold (1887–1948). He was an American ecologist and a hugely influential figure in wildlife conservation and environmental ethics. Leopold grew up in the Midwestern city of Burlington, Iowa, and was the son of German immigrants. From a young age, he spent much of his time exploring the great outdoors and showed an aptitude for observation – a much-needed skill in ecology. Leopold attended the Yale Sheffield Scientific School, and under his yearbook photograph is the comment 'To hell with convention'. I'm not sure whether he wrote this, but it certainly was a theme of his thinking. For example, 'Nonconformity is the highest evolutionary attainment of social animals' is one quote of Leopold that seared into my memory upon reading it.[15] Eventually, Leopold moved to Wisconsin, where he became the first Professor of Wildlife Management. During his time in Wisconsin, he purchased 33 hectares of land in the central 'sand country', which was degraded by years of logging, overgrazing and subsequent wildfires. Leopold spent much of his spare time restoring this landscape and learnt an incredible amount in return. Eventually, Leopold developed what he called the *land ethic*, recognising that 'We abuse land because we regard it as a commodity belonging to us. When we see land as a community to which we belong, we may begin to use it with love and respect.'[16] Leopold's work inspired the development of nature connectedness research.

Nature connectedness defines how connected we feel and how we identify ourselves in relation to the rest of nature. Researchers in this realm consider humans to be deeply embedded within nature – the antithesis to the dominionistic view discussed earlier. Initially, this concept can be strange to relate to and challenging to embrace.

Indeed, I've written in the past about how consciousness – as a biological phenomenon – is steeped in intrinsic complexities. As individual human beings, we show fundamental asymmetries in how we view ourselves and how we view others. We can attribute this partially to our intricately complex emotional and cognitive immersion. Viewing ourselves as a component of nature or deeply embedded within the rest of the living community could also prove to be a challenging feat. Yet, as you're reading this, try to think about your relationship with somebody close to you – for example, your partner or your best friend. You probably (hopefully) feel some kind of deep connection or bond with them – but you may not actively think about this bond and its inherent meaning all of the time. You probably have several things in common and many shared experiences and memories. Indeed, neuroscience has shown that similarities are, on average, greater between two friends than between two people chosen at random. You probably feel a warmth towards them and are comforted by the familiarity of their idiosyncrasies and mannerisms. Your relationship may even feel like it is an integral part of your identity. Hopefully, you think about how your actions affect this person.

There may be several evolutionary explanations for these attributes; for example, genes influence the formation and structures of social ties. But importantly for this discussion, all of these thoughts and feelings can also manifest and be harnessed in a relationship between yourself and the rest of the natural world – the trees, the lakes, the birds, the soil, and dare I say the microbes, but from a collective 'nature' perspective.

People who have higher levels of nature connectedness are more likely to exhibit 'pro-ecological' behaviours.[17] These behaviours include but are not limited to a desire or keenness to conserve and protect nature, to recycle and prevent unnecessary waste, and to donate to wildlife charities. Researchers have also linked nature connectedness to *eudaimonia* – an elaborate Greek word for human flourishing and vitality. There is also a degree of plasticity – that is, the connection is not static – to one's emotional connection to nature, and this is a wonderful point to reflect on.

Why is this so wonderful, I hear you ask? It is wonderful, in my opinion, because this plasticity means that it is possible to enhance one's connectedness to nature. This could open the doors to a suite of profoundly positive health and well-being outcomes and bring us closer to appreciating the unseen world. Nature connectedness could be a fantastic enhancer of health and, perhaps the most remarkable thing of all, it's completely free. Admittedly, writing the last rather insipid line, presenting an advert-esque view of nature connectedness, did make me cringe slightly. However, the very notion of nature being a free wonder-therapy also reminded me of reading Matthew Walker's book *Why We Sleep: The New Science of Sleep and Dreams*.[18]

In one of the opening passages to this book, Matthew refers to a revolutionary, new all-in-one wonder drug that can make you live longer, enhance your memory, protect you from diseases and reduce stress and anxiety. Of course, he refers to the simple (for some, but not others, and I include myself in the latter camp) act of sleeping – more specifically, getting quality sleep and getting enough of it. If a single pill could help cure the diverse suite of maladies listed above, drug companies would undoubtedly make gazillions. Nature connectedness, like quality sleep, could be seen as much more than just a free wonder drug; it could enhance your health and well-being and that of the environment.

The following is an exercise which, at a very fundamental level, allows you to self-assess how connected or 'at one' you feel with the rest of the natural world. It's a simple and concise instrument called the Inclusion of Nature in Self scale (INS).[19] Although not a comprehensive test, the INS has been 'validated' (a process in which the instrument is reviewed to determine whether it measures what it was designed to measure) and used by environmental psychologists in a relatively large number of scientific studies. A person undertaking the INS assessment is asked, 'Please indicate which picture (A–G) best describes your relationship with the natural world.'

Have a go yourself – simply look at each picture below, think about how connected you feel to nature, and choose one of the images. Image A describes a feeling of complete separation between yourself and the natural world, whereas image G describes a sense of wholeness, of being at one or deeply connected with nature.

I feel very connected to nature. However, I can never quite bring myself to select the highest level (G), possibly due to the ravages of modern living.

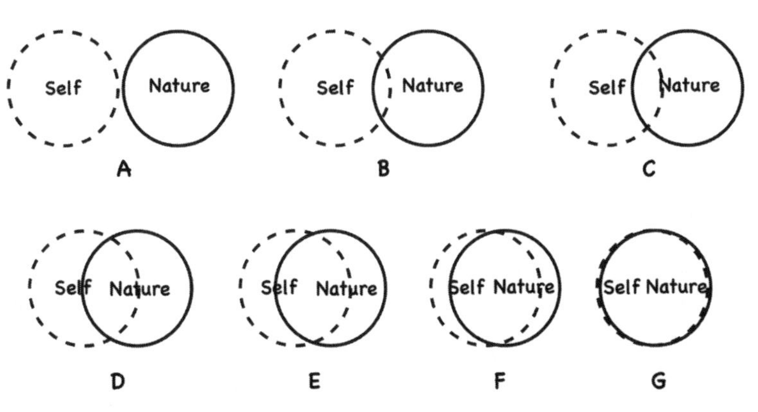

The Inclusion of Nature in Self scale.

Researchers use several comprehensive questionnaires to assess people's nature connectedness. For example, there's the Connectedness to Nature Scale and the Nature Connection Index. The latter was developed by Professor Miles Richardson (a world-leading authority on nature connectedness research) with his team at the University of Derby.[20] Perhaps the most widely used nature connectedness scale in research is the Nature Relatedness 6-item scale (NR-6) developed by Professors Elizabeth Nisbet and John Zelenski.[21]

Again, try it yourself. In the table below, simply tick the box that best describes your feelings for each statement. This is a Likert-style scoring system whereby ticks in the 'Strongly Disagree' box score 1; ticks in the 'Slightly Disagree' box score 2; and it goes all the way up to the 'Strongly Agree' box, which scores 5. Once complete, total your scores. If the total is between 6 and 12, this indicates that you score low on the nature connectedness scale, whereas if the total is between 24 and 30, you score highly on the scale and likely feel very emotionally connected to nature.

How to enhance your nature connectedness

If you scored fairly low on any of the above, fear not! As I said earlier, nature connectedness is not a static concept, and you can enhance it through various nature engagement activities.

	Strongly disagree	Slightly disagree	Neutral	Slightly agree	Strongly agree
I feel very connected to all living things and the Earth					
I always think about how my actions affect the environment					
My relationship with nature is an important part of who I am					
My connection to nature and the environment is a part of my spirituality					
My ideal holiday/vacation spot would be a remote, wilderness area					
I take notice of wildlife wherever I am					

The Nature Relatedness 6-item scale.

Take moments to mindfully notice nature throughout your day and write down a few good things that you find pleasing. This could be the song of a robin, the warm colours of autumn leaves, or vivid green moss that grows in the cracks of a wall. Try to notice nature wherever you are. If you can access a park or woodland, try to spend some quiet time reflecting by using all your senses – notice the sights, sounds and smells of nature, learn how to go foraging, and this will give your connectedness a boost and bring you closer to our invisible friends.

Sometimes, however, it can be hard to access nature. If this is the case, try bringing nature to you. Bring plants into your house. Try growing various pot herbs for that extra olfactory and gustatory interaction. If you have a garden or a balcony, consider attracting wildlife – bird feeders, log piles or 'bug hotels', and a variety of wildflowers, can help enormously. Nature connectedness is about meaning: a spiritual, emotional and cultural connection with nature, and mindfulness. If you can take time to notice nature mindfully, be inspired by nature, and do this regularly, you can enhance your nature connectedness in the long term. Suppose you can do these things whilst thinking about and appreciating how the nature you *can* see intimately depends on the nature you *can't* see. In this case, you may eventually find the experience humbling and transformative.

My aim for writing this chapter was to help promote a closer connection to nature and stimulate a new appreciation for the unseen members of our biotic community. Nonetheless, in COVID-19 we have an example of a single microscopic species, the SARS-CoV-2 virus, which has potentially changed our exposure to nature via biological, social and political transmission pathways. If we think about it, the pandemic has not only affected us directly through the nasty infections, but has also led to enormous changes to our lives that are socially and politically transmitted – including lockdowns and other social restriction measures.

To what extent our exposure to natural environments has changed due to the COVID-19 outbreak was the topic of another study I was involved in during my PhD. In this study,[22] we found that during the COVID-19 lockdowns of 2020 in the UK, people increased the amount of time they spent in nature. They visited for longer and visited more often. Most people said they travelled to these local 'green' and 'blue' spaces (i.e. terrestrial and aquatic habitats, respectively) for health and well-being reasons. This highlights the importance of conserving and restoring natural environments to maintain resilient societies and promote the health of our planet. However, other studies have revealed considerable disparities in people's ability to access quality and vibrant natural environments, as discussed in the social equity chapter (Chapter 4). This means that many people who endure these pandemic lockdowns are meanwhile unable to access a fundamental source of health creation – nature! This will likely have a significant impact on their ability to foster a deeper connection with the natural world, thereby stifling pro-ecological behaviours and an appreciation for the unseen. As I highlighted above, however, we can often find nature wherever we are if we seek it mindfully.

The COVID-19 virus has had an unfathomable impact on our health, economic systems and social lives across the globe. It has highlighted the fragility of humanity and the deep interconnectedness of the biosphere and sociosphere. Yet many microbial species are vital to our health, and we must not lose sight of this. Resisting germophobia (the fear of all microbes and 'dirt') and supporting the restoration of ecosystems and their microbiomes is essential to our health and the planet. The act of fostering deep respect for ecosystems has never been needed more. Opportunities to enhance our health and the rest of the natural world can arise through a mindful journey of learning about our identity and our connections within nature. A feeling of warmth and comfort through familiarity and judgement-free interactions can be developed by re-establishing our own ancient connections with the land and the wondrous organisms who share it with us. You may already feel this, but if not, you could take action to reconnect – and eventually feel as though nature and those remarkable unseen interactions are an essential part of your identity. A beacon of strength, awe and belonging.

The environment, microbes, human health and nature connectedness. They're all interrelated, and harmony between them is vital –now more than ever.

CONCLUSION

Great and Small

Can we change the prevailing negative perception of microbes, the invisible biodiversity that shapes our lives and the world around us? I began writing this book to draw together concepts and stories that illuminate the vital roles of microbes in the rich tapestry of life. As the book comes to a close, I still believe the answer is yes. We can, and indeed must, change this perception. By doing so, we can redefine our relationship with nature and curate a new enduring connection and ethic that benefits future generations – of humans and other-than-humans alike.

The invisible and intangible nature of microbes poses a considerable challenge to our perception, empathy and appreciation. I hope the topics explored in this book offer some starting points to inspire you to think about how life interconnects, with microbes being the glue that holds it all together. Knowing this can be humbling, but it can also teach us lessons; lessons about cooperation and diversity – both of which are needed in society, arguably more now than ever as we face immense social, political and environmental challenges globally.

Whilst I was writing the book, something unexpected happened. I was contacted unexpectedly by Professor Kenneth Timmis, a Fellow of the Royal Society who works at the Technical University of Braunschweig in Germany. He asked me if I'd be willing to contribute to a project called 'The urgent need for microbiology literacy in society'. As you can imagine, my curiosity was immediately piqued, and I was honoured to be invited. In short, the project aims to curate a freely adaptable selection of generic lessons that can shape school curricula for children in diverse teaching settings.

The goal was to reach teachers interested in learning about the wonderful world of microbes. This sounded fantastic, and it was a perfect opportunity to inspire children by engaging them in the microbial ecology concepts that touch and shape their lives. Projects like these instil a great sense of hope. As discussed in this book, early life is the vital window of opportunity to ensure biological and cognitive connections are formed between humans and the microbial world.

In this book, and relevant to early life, we delved into the vivid world of immunology with Professor Graham Rook and discussed the biodiversity hypothesis. The effects of exposure to rich microbial diversity from a young age are far from evanescent; indeed, such exposures are imperative for a robust immune system later in life. They may also be important in shaping our minds and behaviour, as microbes chatter away to the brain and vice versa through the two-way communication highways in animal bodies.

As we discovered with Dr Sue Ishaq, we can also view microbes as a facet of social equity. Our invisible friends keep us healthy. Yet where we grow up, what foods we have access to, the quality of the natural environments around us, and our socioeconomic status all affect our exposure to microbes – both the friends and the foes. We know deprivation differs enormously across the world. Life expectancy can even vary by nearly 20 years within cities and adjacent neighbourhoods (i.e. the Glasgow Effect). The available opportunities to sustain a healthy microbial ecosystem both inside our bodies and out likely have a role in determining these stark statistics. Therefore, we must ensure that access to health-promoting microbial friends is equally distributed.

Our landscapes and microbioscapes are changing. Bacteria continuously swap among themselves vast arrays of antimicrobial-resistance genes, far across our terrestrial and aquatic ecosystems. This poses a threat to our ability to control diseases. Scientists are developing phage therapies to counteract this growing issue; one of our most salient group of friends might turn out to be viruses – a strange thing to utter in the COVID-19 era.

I am not convinced that we should always attribute some behaviours or traits to mere 'personality', or variations in such behaviours to 'moods' because, from a holobiont perspective, microbes are likely tinkering behind the scenes. I agree with Professor Robert Sapolsky that we should deeply consider the long list of biological, social and cultural triggers that lead to a given behaviour and evaluate the orchestral role of the brain, body and the environment (including microbes) in addition to reductionist approaches. In the Lovebug Effect chapter, Dr Martin Breed and I pondered whether microbes could influence our decisions to spend time in certain environments that may benefit our body's microbial ecosystem. And in the holobiont blindspot chapter, I discussed the possibility that our microbiome affects our perceptions and actions, or 'System 1 thinking', by regulating our impulses. Should we consider this when debating the notions of free will and determinism? These are fascinating questions that transcend the boundaries of current knowledge and are worthy of further enquiry owing to their potentially important social ramifications.

Microbes influence every corner of the world and every second of our lives. The microbiology revolution that occurred during the nineteenth century spawned a plethora of innovative and instrumental applications that benefit society. We use them to influence the flavours and textures of some of our most wonderful foods and drinks, although we did this way before we knew they existed. Now microbes are influencing forensics, space exploration and the architectural skins of our cities. The last of these is a particularly exciting application, as we need to live more sustainably in our dense urban jungles. Can we increase the biodiversity of our buildings' desolate surfaces so that humans and wildlife can both benefit? And can we apply microbial ecology principles in our urban environments to help ecosystems flourish? I'm excited to be working on some of these research questions alongside the wonderful colleagues I interviewed whilst writing this book.

We have microbes to thank for our foods, nutrient cycling, disease control, climate stability and many other things. Restoring our ecosystems – our abodes – is crucial to prevent further biodiversity loss and catastrophic climate change. A greater focus on microbes' roles in ecosystem restoration will undoubtedly be vital.

We know that by working with nature instead of overpowering it, we can foster deep reciprocity and generate mutually advantageous outcomes for people and the rest of nature – this is what Professor Robin Wall Kimmerer calls a 'gift economy'. One important social strategy to adopt alongside ecosystem restoration is enhancing our nature connectedness – our emotional, cognitive and experiential connection with nature. I feel we can enrich this phenomenon by taking moments to think about and marvel at the wondrous microscopic activity beneath our feet, within the tree bark and in passing birds, and knowing that our limited visual acuity can no longer restrict our appreciation for our invisible friends.

As I write the conclusion to this book, I have just heard the news that Professor E.O. Wilson – the great American evolutionary biologist – has sadly passed away. As mentioned in the previous chapters, Wilson's work on biodiversity, the biophilia hypothesis and the interconnectedness of life has been a real inspiration, shaping my research and world view. Wilson suffered an injury to his eye as a child during a fishing accident. His long-range vision was impaired. As a result, he studied the smaller creatures he could view up close, such as ants and other tiny animals. His work on the diminutive and less-appreciated life-forms sparked a lifelong passion, and he knew acutely, as Louis Pasteur said, that 'the role of the infinitely small in nature is infinitely great'.

APPENDIX

Microbes 101

For those of you who'd like to know a little more about the different types of microbes that roam the planet, here are a few pages describing bacteria, archaea, viruses, fungi, algae and protozoa.

Bacteria

When the Dutch scientist Antonie van Leeuwenhoek built a rudimentary microscope in 1671, he began observing the micro-organisms he found in water, skin, hair and even his faeces. Some of these microbes were bacteria, and he was the first person ever to see them. Although he endured an onslaught of scepticism for much of his career, his microscopy work greatly influenced science in the following centuries. Indeed, he is now widely regarded as the 'father of microbiology'.

Van Leeuwenhoek called the bacteria and the other microbes he observed 'animalcules'. Many years passed before bacteria were studied in detail, mainly because other scientists could not replicate the quality of van Leeuwenhoek's microscopes. Many influential scientists also ridiculed his writing style, which probably didn't help. Nevertheless, van Leeuwenhoek had single-handedly 'discovered' several domains of life without the help of these influential scientists. It is bewildering to think that before this fateful event, humans had lived, not just alongside, but deeply interlinked with these domains of life for our entire existence without appreciating their sheer beauty and diversity – a multitude of unseen and unsung friends and foes alike. The human world changed forever when this invisible cosmos was finally revealed. Our understanding of bacteria (which are single-celled organisms) and other microbes, and of the roles they play in our ecosystems and our lives, led to revolution upon revolution.

The word bacterium is a Latinised form of the Ancient Greek word *baktērion*, meaning 'staff' or 'cane' – because the first bacteria discovered were rod-shaped. Sometime between two and four billion years ago, bacteria evolved on Earth. Some of the earliest were cyanobacteria, which dwelled in the primordial oceans. Before they arrived on the scene, Earth's atmospheric oxygen was only 1%. In the two billion years that followed, cyanobacteria often formed into tightly packed deposits called *stromatolites*. The cyanobacteria were and still are photosynthetic and, as such, they would spend aeons pumping oxygen into the oceans. Once the oceans were saturated, the oxygen escaped into the atmosphere. Over time, the oxygenated atmosphere allowed many other life-forms to evolve and flourish. It is not an overstatement to say that we have these stromatolite bacteria to thank for our very existence. Some years ago, I had the pleasure of visiting a massive colony of stromatolites in Western Australia. They are living fossils, comprising dense mattresses of bacteria and carbonated deposits protruding from the coastline; still living, and, in evolutionary terms, some of the oldest life-forms on Earth.

We share the planet with millions of bacterial species, and 99% have yet to be described. They primarily reproduce by a process called *binary fission*, whereby a single cell divides into two. This is a form of asexual reproduction. Bacteria are found in almost any environment you can think of, from the hot springs of Yellowstone National Park to the lakes buried under nearly a kilometre of ice in Antarctica. They are essential to the integrity of ecosystems and play vital roles in plant health, plant communication, animal health, nutrient cycling and climate regulation. For many organisms, bacteria are food, and for others, they provide food in exchange for resources they need – forming a mutual, symbiotic relationship. Although bacteria may seem like simple prokaryotes, they have extraordinary capabilities. They can communicate with each other in a process called quorum sensing. They can then form biofilms to mount group-based responses, generate electricity, degrade pollution, stimulate rain and achieve many other spectacular feats.

Archaea

One of the major dogmas of biology was overturned in 1977. For many years, scientists had classified all life on Earth into two domains – Prokaryota and Eukaryota. Then along came American Professor Carl Woese, who blew this assumption out of the water. Woese discovered that within the Prokaryota domain, two distinct groups of organisms existed that were only distantly related to each other.[1] This discovery is now widely accepted, and, as a result, the third domain of life – Archaea – was proposed. The new classification scheme has three distinct groups of life: Eukarya, Bacteria and Archaea.

Like bacteria, archaea are prokaryotic and unicellular. They also undergo binary fission to reproduce. Archaeal cell membranes have a different structure and configuration to bacterial cell membranes, which can be helpful for protection in so-called extreme environments (they're extreme from a human perspective, at least). Scientists first found many archaea in these harsh environments. The conditions of these environments were often highly acidic or very hot. To reflect this, the microbes living in these environments were given the title *extremophiles*. Nonetheless, we now know that archaea exist and thrive in a wide range of environments, from soils to oceans and even our bodies. They are vital players in many ecosystems.

To identify this third domain of life, Woese and his team used genomic methods that were cutting-edge at the time. They compared the gene sequences of a particular molecule called ribosomal RNA, which has a central role in the functioning of cells. Woese's development of this technique revolutionised ecology, and you could argue that it was responsible for the advent of *microbial ecology* as a field of research. By collecting samples from an ecosystem and sequencing a particular gene – the ribosomal RNA gene – researchers were able to sidestep the laborious and often unsuccessful task of culturing microbes from the environment in the lab. In the mid-1990s, Woese and his team published the first complete genome structure of the archaeon *Methanococcus jannaschii*. They concluded that archaea are more closely related

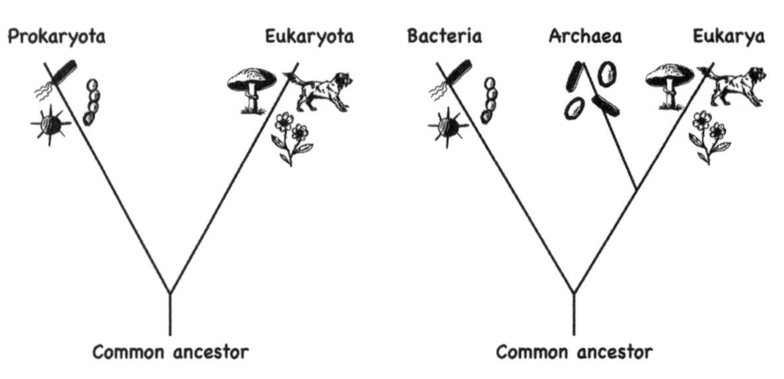

The old and the new domains on the tree of life.

to plants and animals – including humans – than to bacteria. Woese said, 'archaea are related to us, to the eukaryotes; they are descendants of the microorganisms that gave rise to the eukaryotic cell billions of years ago'.[2] Genomic researchers have since found more similarities between archaea and eukaryotes than between archaea and bacteria. This is astonishing, especially as we once considered archaea to be bacteria; we even once called them archaebacteria. Many call Woese the 'man who rewrote the tree of life' due to his game-changing work on archaea.

Viruses

The COVID-19 pandemic has been a ubiquitous backdrop to our lives in the last couple of years. Consequently, by watching the news and social media, most of us have inadvertently received an elementary education in virology. We know that many viruses are highly contagious and can spread from person to person through the air; we know they are parasitic – using the host's cells to multiply and disperse. We know they are microscopic and can be round and spiky in shape. We know they can make us severely ill and even lead to death.

There's no doubt that some viruses, like that responsible for COVID-19 (SARS-CoV-2), are not our invisible friends. Yet viruses are among the most diverse and abundant entities on

Earth. Along with this diversity comes legions of functional roles that are integral to our ecosystems – and it's safe to say that many are, in fact, our friends.

Take a moment to visualise the universe. It's impossible, right? Some estimate the diameter of the observable universe to be 93 billion light years. A single light year is 9.46 trillion kilometres. So that's somewhere in the region of 880 billion trillion kilometres in diameter – or 880 'yottametres'. Scientists estimate there are ten billion times more viral particles in just the oceans of our tiny blue space dot (Earth) than the estimated number of stars in the universe. The total mass of viruses on our planet is equivalent to 75 million blue whales. These are mind-blowing numbers, and they serve to remind us of the profound, invisible and multifarious cosmos in our oceans, in the air around us, and in the soil beneath our feet. Only just over 9,000 named virus species are listed by the International Committee on Taxonomy of Viruses. A surge in virus research following the COVID-19 pandemic led to over 1,000 new species added to the list in 2020 alone.[3]

Many do not consider viruses to be living entities. This, of course, hinges on a person's definition of 'living'. Viruses cannot replicate in the absence of a host, and they essentially hijack the host's cellular machinery to do so. Some would say this is enough to stick viruses in the bucket labelled 'non-living'. On the other hand, is any living entity or organism entirely self-supporting? Many organisms or 'microparasites' are intimately dependent on the physiology, activity and resources provided by the host's cells. Moreover, humans, and all other living entities, are dependent on the ecosystems around them – and as has been discussed in this book, our survival is reliant on the trillions of microbes that call the human body home.

Whether or not we consider viruses living, they play critical roles in ecosystem processes and human health. Through host infections, viruses help maintain the biological equilibrium in our ecosystems. Some viruses, known as bacteriophages or phages, prey on bacteria, thereby controlling opportunists and pathogenic species that might otherwise cause severe diseases in plants and animals. They also play a role in transporting genes between organisms, and thus are essential to the diversity of life on Earth.

It is not fully understood how viruses evolved, and some scientists believe there isn't a single common ancestor. As science journalist Amber Dance recently pointed out, this means 'viruses probably arose several times in the history of life on Earth'.[4] There is also debate over whether viruses evolved from independent genetic material at the dawn of time, or whether they are essentially stripped-down versions of more 'complex' organisms – embracing the minimalist way of life.

Fungi

Take a walk through a woodland and steer your eyes towards the soil, the trees and the rotting logs on the floor. Or stroll across a grassland that hasn't been extensively managed, and eventually you'll notice the fruiting bodies of filamentous fungi. A great range of species exists – again, in seemingly inconceivable numbers across the planet. These are the macroscopic or visible fungi, from the orange funnel-shaped and delicious chanterelles to the pale, hemispherical and deadly death cap. However, a rich assemblage of invisible fungi also exists, many of them living inside us, and even more in the ecosystems we depend upon. There are yeasts, rusts, smuts, moulds and mildews. They play critical roles in these ecosystems, from nutrient cycling to plant communication.

We call fungi, like animals, *heterotrophic* (pronounced 'hetter-oh-troh-fick') organisms, meaning they cannot produce their own food and must acquire it from other organisms such as plants and animals. Even though fungi were considered plants as recently as the 1960s, the genetic composition of fungi is more akin to that of animals. Many fungi live cryptic lifestyles, so we rarely see more than a few species. They have a fundamental role in nutrient cycling and decomposing organic matter. Many are mutualistic symbionts – providing nutrition, pathogen protection and a means of communication in plants (via mycorrhizal networks) – and some are parasitic, with critical roles in maintaining a biological equilibrium, much like viruses. Around 150,000 species have been described, but some estimate that up to four million species call Earth their home. About 1,500 species of newly discovered fungi are described each year.

The word fungus comes from the Latin word for mushroom. This, in turn, is derived from the Greek word *sphongos*, meaning 'sponge', which refers to the organism's often-spongey structure. While most fungi grow as threadlike structures called *hyphae*, many microscopic species are single-celled like bacteria. These grow and reproduce through a phenomenon called budding. A bud develops on the cell's surface, and the parent cell's nucleus divides. At which point, one of the 'daughter' nuclei remains in the parent cell, while the other migrates into the bud. The buds eventually pinch off and become separate fungal individuals. This is a form of asexual reproduction, but some fungi can also undergo sexual reproduction, whereby 'haploid' spores with a single set of unpaired chromosomes are produced. These can then recombine with other haploid spores to produce a 'diploid' cell with two complete sets of chromosomes, one from each parent.

A team of palaeobiologists recently travelled to the dramatic cliffs of the Canadian Arctic in a region called the Grassy Bay Formation. They found some tiny fossils believed to be ancient fungi. They think it might be the oldest record of a fungus ever discovered. The team named the fungus *Ourasphaira giraldae*.[5] It was found in billion-year-old rock. Previous molecular analyses had dated the origin of fungi to around a billion years ago, so the recent finding supports this claim.

The immense diversity of fungi means they occupy many ecological niches and carry out a plethora of ecological functions. In humans, fungi are essential members of the microbial communities residing in and on our bodies. The human 'mycobiome' (fungal communities in our bodies) is relatively underexplored. While some species are harmful and can cause fungal infections, many others likely play a positive role in our internal ecosystem.

Algae

The word alga (algae is the plural) comes directly from the Latin *alga*, meaning seaweed. Most algae live in freshwater or seawater. Many comprise large, joined-up cells like seaweed, but some are free-floating – also known as 'planktonic'. Others live in tight-knit mutualistic relationships with fungi and bacteria to form the lichens that decorate tree trunks, rocks and walls. All algae contain chlorophyll pigments, making their food through photosynthesis. As in plants, the chlorophyll sits neatly inside specialised cells called chloroplasts, and it gives some algae their characteristic green appearance.

Diatoms are algae that float around in the seas and lakes. In a diatom's cell wall is a rigid substance called silica. When the diatoms die, they sink to the bottom of the sea. Over time, the immense pressure of the water forces the silica molecules together to form a thick, rocky layer across the seabed. This is then mined by humans and used in various products, including toothpaste and polishes. In aquatic ecosystems, algae play a foundational role in the food web and provide nutrients for many other organisms. Because algae form huge blooms and are photosynthetic, they produce vast amounts of oxygen in our atmosphere. Rainforests are often called the 'lungs of the Earth', yet one out of every two breaths we take is filled with oxygen that phytoplankton produce – they account for 50% of the Earth's oxygen production.[6] Now that is what I call an invisible friend.

Some marine algae are indeed microscopic. Millions of species float across the great pelagic regions of our oceans. Their body shapes are also diverse. Some have tiny spines that help maintain buoyancy, while others have tail-like appendages called flagella, which help them navigate through the water. We also find algae in soils. They have a role in introducing organic matter into the soil and excreting sugary substances that increase stable soil aggregation – this is important for reducing soil degradation and retaining water. People often call cyanobacteria (a type of bacteria) blue-green algae because when they clump together they resemble algae. Cyanobacteria are photosynthetic, but unlike algae, which are eukaryotes, they are prokaryotes – in other words, they lack a true nucleus and membrane-bound organelles.

Humans use algae for biofuel production, animal feed, crop fertiliser and a heap of other things. In Chapter 11, I discussed more about the innovative ways algae support our ecosystems and our health.

Because algae are photosynthetic organisms, they are unlikely to live in the deep, dark crevices of the human body for very long. However, algae provide a range of beneficial nutrients and chemicals that humans consume: omega oils, vitamins, proteins, antioxidants and carbohydrates are a few. A recent study showed that an alga called *Chlamydomonas reinhardtii* helped improve the symptoms of irritable bowel syndrome, including reducing diarrhoea and bloating.[7] So, even if they're not a permanent resident in our microbiome, algae could still benefit our microbial ecosystem.

Protozoa

Antonie van Leeuwenhoek, the bacteria-discovering Dutch microscope maker, was also the first to see protozoans. Van Leeuwenhoek first described protozoans as 'animalcules', just as he did with bacteria. The word protozoa comes from the Greek *proto* ('first') and *zōon* ('animal'). This reflects their animal-like behaviour. However, despite their name and behaviour, they are single-celled eukaryotes.

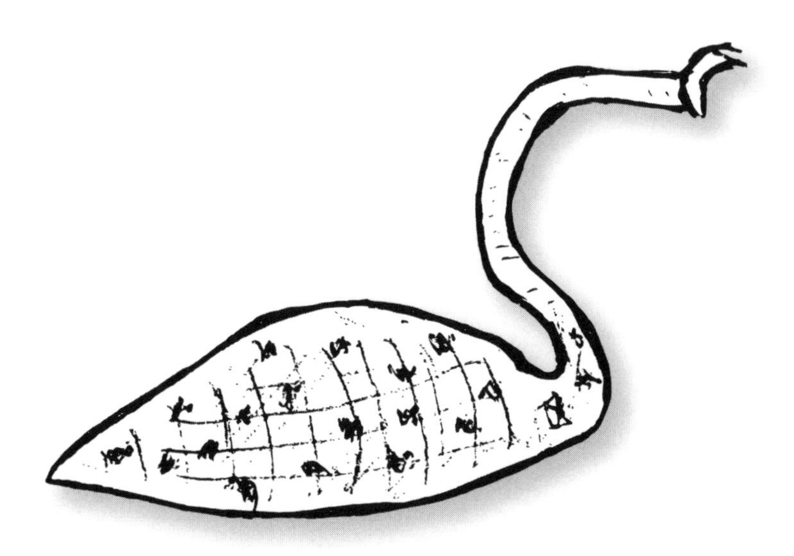

Lacrymaria olor, or 'tear of a swan'.

Scientists have described more than 50,000 protozoan species. We find them in almost every known habitat on Earth. Humans have protozoa living on or in their bodies. We've had a long co-evolutionary history with protozoa but know relatively little about how they affect our human ecosystem. Because we have this co-evolutionary history, they likely provide significant health benefits. Like the other microbes, protozoa play essential roles in our broader ecosystems. They provide a regulatory role by consuming bacteria, and they mineralise nutrients, making them available for plants and other soil dwellers.

Just like phytoplankton, protozoa come in all shapes and sizes. There's the trumpet-esque *Stentor muelleri*, which feeds on bacteria by pumping water into their cells. There's also the slender-necked *Lacrymaria olor* – which means 'tear of a swan'. This is an excellent description of its appearance, and its ability to extend its 'head' outwards and turn it in many directions.

Glossary

aerobiome: The network of all microbial communities (and their genetic material) within a given airspace.

allergen: A substance that can cause an allergic reaction.

animal model: A non-human species used in biomedical research because it can mimic elements of a human biological process or disease.

archaea: Single-celled microorganisms with a structure similar to bacteria. They are evolutionarily distinct from bacteria and eukaryotes. They form the third domain of life and are often found in extreme environments, but not always! Moreover, 'extreme' is a relative term – an environment that is extreme to humans is not for other organisms.

bacteriophage: A type of virus that infects bacteria, also known as 'phages'.

biodiversity hypothesis: The hypothesis that contact with natural environments enriches the human microbiome, promotes immune balance and protects from allergy and inflammatory disorders.[1]

bioinformatics: The science and computational study of biological information. An interdisciplinary field that develops methods and tools to understand often large and complex biological data.

biophilia: A hypothesis that proposes humans possess an innate affinity to connect with other forms of life. Edward O. Wilson introduced and popularised this hypothesis in his book *Biophilia* (1984).

biophilic drive: A process which describes the motivation behind an individual's desire to seek out natural environments and other life-forms.

chloroplast: An organelle within the cells of plants and certain algae that is the site of photosynthesis.

cognitive bias: A systematic error in judgement or deviation from rationality – primarily to save our brains time/energy.

decibel (dB): A sound intensity, also known as amplitude, measured on a logarithmic scale.

DNA sequencing: The process of reading and interpreting the nucleic acid sequence, i.e. the order of the building blocks of DNA.

dysbiosis: A term used to describe an imbalance or maladaptation in a microbiome (collection of microbial communities in a given environment), typically with adverse effects on animal health.

ecological restoration: The process of assisting the recovery of an ecosystem that has been degraded, damaged or destroyed. Restoration ecology is the corresponding scientific discipline.

endosphere/endophytes: The area inside a plant or the 'internal world'; the microbes in this area are known as endophytes.

eubiosis: Microbial balance in the body.

eukaryote: Organisms such as plants, animals and fungi, that have a membrane-bound nucleus and organelles.

forest bathing: A Japanese practice (*shinrin-yoku* 森林浴) of immersing oneself in a forest environment – a method of being calm amongst trees for a well-being benefit.

germ: A colloquial term for microbe, most often used to describe pathogens.

germophobia: The pathological fear of or aversion towards microorganisms and dirt.

green infrastructure: 'Strategically planned network of natural and semi-natural areas with other environmental features designed and managed to deliver a wide range of ecosystem services in both rural and urban settings'.[2]

holobiont: A term first coined by Margulis (1990), defined as a 'biomolecular network composed of the host plus its associated microbes [i.e. the Holobiont], and their collective genomes [which] forge a Hologenome'.[3]

holobiont blindspot: Failing to consider the role of host–microbiome (collectively termed a 'holobiont') interactions in a given behaviour, which may underpin a potentially important cognitive bias.

hologenome: Collective genomes of holobionts.

horizontal gene transfer: Also known as lateral gene transfer; the movement of genetic material between unicellular and multicellular organisms by means other than vertical transmission of DNA from parent to offspring.

Lovebug Effect: An evolutionary model to describe microbially mediated nature affinity.

macrophage: A type of white blood cell of the immune system that engulfs and digests invading cells.

metaorganism: A host and its complete associated microbial community.

microbe: Also known as microorganism. Microscopic organisms that exist as unicellular, multicellular or cell clusters. Examples include bacteria, fungi, viruses, archaea, protozoa and algae.

microbiota–gut–brain axis: A bi-directional communication link between the microbiota, the gut and the brain.

microbiome: The entire collection of microorganisms (and their genetic material) in a given environment and its ecological theatre of activity.

microbiome-inspired green infrastructure (MIGI): Natural infrastructure that is restored and/or designed and managed to promote reciprocal (often health-promoting) relations between humans and environmental microbiomes, whilst sustaining microbially mediated ecosystem functionality and resilience.

microglia: A type of immune cell located throughout the brain and spinal cord – they are the brain's resident macrophages.

mitochondrion (plural: mitochondria): The tiny organelles in our cells that provide cellular energy.

nature connectedness: One's emotional, cognitive and experiential connection with the rest of the natural world.

old friends hypothesis: An update by Rook et al. (2003) on the hygiene hypothesis,[4] suggesting that because of our long evolutionary association with certain microorganisms, they are recognised by the innate immune system as harmless – or in some cases, treated as 'friends' because they are needed for regulation.

phage: See *bacteriophage*.

phyllosphere: The total above-ground portions of plants (the leaf world) and their habitat for microbes.

Prebiotic: a non-digestible food item that promotes the growth of beneficial microbes.

prokaryote: Organisms that do not have a membrane-bound nucleus, like bacteria.

protozoa: Single-celled eukaryotes including amoebas, flagellates, ciliates, sporozoans and others. They are treated as phyla belonging to the kingdom Protista.

rhizosphere: The region of soil that is directly influenced by root secretions and associated soil microorganisms, known as the root microbiome.

short-chain fatty acid (SCFA): The main metabolites (metabolic by-products) with fewer than six carbon atoms, released by microbes in the gut. They are often important in human health.

Symbiocene: A philosophical term used to describe a proposed geological epoch defined by a mutually advantageous relationship between all living beings (particularly between humans and the rest of nature).

symbiont: an organism living in close association with another (often denoting the smaller organism in the partnership).

System 1 thinking: A conceptual branch of cognition characterised by 'fast and automatic thinking' – popularised by Daniel Kahneman.[5]

Notes

Introduction: A Hidden World

1. Around 60 million years ago, there were many active volcanoes across the UK. Remnants of these volcanoes can be seen today. See the following link for more information: www.countryfile.com/go-outdoors/historic-places/britains-most-amazing-extinct-volcanoes/.
2. Bear and Thomas, 1964. Australians Isabel Joy Bear and R.G. Thomas coined the term *petrichor*. The authors described it as 'the smell derived from an oil exuded by certain plants during the dry period'.
3. John Innes Centre, 2020. Sharks can detect a tiny amount of blood from hundreds of metres away.
4. Becher et al., 2020. Springtails eating *Streptomyces* and dispersing their spores is analogous to birds eating the fruits of plants and dispersing their seeds.
5. Bolourian and Mojtahedi, 2018. Inflammation plays a crucial role in healing, but chronic inflammation increases the risk of various diseases, including some cancers.
6. Kannadan, 2018. Miasmas were thought to be toxic emanations, from putrefying carcasses and vegetation and other particles.
7. Microbiology by numbers, 2011; Locey and Lennon, 2016. Only 1,400 known human pathogens exist and new estimates suggest one trillion microbial species live on Earth.
8. Timmis et al., 2019. Avoidance of biodiversity could be contributing to our reduced connectedness to nature and to an increase in immune-related disorders such as asthma, arthritis and diabetes.
9. Gilbert and Neufeld, 2014. Mitochondria and chloroplasts are the powerhouses of animal and plant cells, respectively. And they are thought to have evolved from microbes millions of years ago.

Chapter 1: The Microbiome

1. Locey and Lennon, 2016. These figures change as new technology becomes available. Often we underestimate and the numbers increase.
2. Atchison et al., 1965. R.W. Atchison was a world-leading virologist.

3. Schulz, 2002. *Thiomargarita namibiensis* is now the second-largest known bacterium in the world. At the time of writing these notes, scientists have just described an even larger bacterium – *Thiomargarita magnifica*, living in the Caribbean! It's thought to be around 1 cm in diameter – so visible to the naked eye and calling into question the definition of a 'microbe'.
4. Shkoporov et al. 2022.
5. Ma and Mei, 2022.
6. Breitbart et al. 2018.
7. Sieradzki et al. 2019.
8. Berg et al., 2020. The definition of 'microbiome' differs between disciplines and even between people within disciplines. A biome can be defined as a large, naturally occurring community of flora and fauna occupying a major habitat, e.g. forest or tundra. Therefore, you can view a microbiome as a small one of these – but for microbes.
9. Yong, 2016. Ed Yong is a science journalist at *The Atlantic* and has also interviewed microbiome scientists including Professor Rob Knight, a pioneer of many microbiome computational tools.
10. Sagan, 1967. Lynn Margulis wrote a paper in 1967 (when she was Lynn Sagan) developing and popularising the endosymbiosis concept – the merging of two lineages through symbiosis went against the modern synthesis of natural selection.
11. Fierer et al. 2008. These figures are applicable to a healthy human body. Someone with dysbiosis (an deleterious imbalance in their microbiome) may have far fewer species.

Chapter 2: Rekindling Old Friendships in New Landscapes

1. Researchers recently found fossils of a badger-like mammal that lived 66 million years ago in Madagascar. It is known as the 'crazy beast': www.reuters.com/article/us-science-crazybeast-idUSKBN22B2BK/.
2. Rook, 1968. Arthur Rook wrote important articles on keratoacanthoma and blistering skin diseases.
3. For more information on electrophoresis: www.genome.gov/genetics-glossary/Electrophoresis/.
4. About 10% of latent tuberculosis infections progress to active disease, which, if left untreated, kills about 50% of those affected.
5. Lowry et al., 2016. Immunising the mice with *M. vaccae* blocked allergic sensitisation.
6. Strachan et al., 1989. The paper on the 'hygiene hypothesis' in the late 1980s. Strachan worked on several other studies relating housing condition to health.
7. Thomson et al., 2010.

8. This was due to several studies reporting correlations between cleaning agents and allergies.
9. Rook and Bloomfield, 2021.
10. Rook et al., 2005.
11. Hesselmar et al., 2018. The recent study 'Pet-keeping in early life reduces the risk of allergy in a dose-dependent fashion'.
12. Davis, 2020. An excellent overview of the immune system was published by Daniel Davis and is called *The Beautiful Cure*. He recounts how the field of immunology represents a health revolution.
13. Rodriguez et al., 2012.
14. Leo and Campos, 2020. Consumption of ultra-processed foods increases the risk for metabolic diseases via changes in the gut microbiome.
15. Stein et al., 2016; Honeker et al., 2019. The Amish population's lower prevalence and severity of inflammatory diseases was linked to microbiota via house dust and faecal samples given to germ-free mice. In future, the study sample sizes need to be increased and with human subjects.
16. Haahtela et al., 2015. This paper discusses the 'hunt for the origin of allergy – comparing the Finnish and Russian Karelia'.
17. Roslund et al., 2020. Marja's study. The Finnish are leading the way on environment-microbiome-immune system studies.
18. Nyanza et al., 2014. There are also reports of Indigenous Peoples eating clay in the Andes and other parts of the world.
19. Ngure et al., 2013. The researchers conducted observations of 23 infants for 130 hours.
20. Troyer, 1984. Green iguanas are but one of many species known to consume parental faeces, potentially due to their rich microbial content.
21. Osawa et al., 1993. This pap-eating behaviour prepares koalas for the dietary transition from milk to the tannin-laden eucalyptus leaves.

Chapter 3: Antibiotic-Resistant Landscapes

1. Greenland sharks can live for around 400 years, making them the longest-lived vertebrate on Earth: www.bbc.co.uk/news/science-environment-37047168.
2. Waglechner et al., 2019. The beginning of antimicrobial resistance: www.sciencedaily.com/releases/2019/08/190812172333.htm.
3. Peterson and Kaur, 2018.
4. Larsen et al., 2022. Their results suggested that a form of MRSA emerged in the pre-antibiotic era as a co-evolutionary adaptation of *S. aureus* to hedgehogs infected with dermatophytes (fungal pathogens in the skin).

5. Interview with Teruaki Nakatsuji, PhD, project scientist in the Department of Dermatology at UC San Diego School of Medicine and first author of the *S. aureus* study: www.sciencedaily.com/releases/2017/02/170222150302.htm.
6. Aminov, 2010. The Aminov paper is a very interesting read.
7. Flanders and Saint, 2014.
8. Aminov, 2010; Rahman et al., 2021. Today, antibiotic residues can be found in milk across the world – including tetracyclines and fluoroquinolones.
9. Article on these statistics with linked studies: www.theguardian.com/environment/2018/jul/19/rising-global-meat-consumption-will-devastate-environment.
10. Kirchhelle, 2018. This is a fascinating paper.
11. Van Boeckel et al. 2015.
12. www.theriverstrust.org/about-us/news/sewage-update-2020-spill-data-added-to-our-map.
13. Ernest Hankin suggested that the presence of antibiotics in the waters of African rivers may have had a role in restricting cholera outbreaks. He studied the Ganges and Jumna rivers.
14. Wittebole et al., 2014. This is a fantastic paper on the early marketing of phages as antibiotics.
15. Chen et al., 2019. Their results suggest higher microbial diversity can act as a barrier to resist the spread of antibiotic resistance.
16. Li et al. 2022.
17. Zhu et al., 2021a. This research group does a lot of great work in the soil microbiome world.

Chapter 4: Microbes and Social Equity

1. A study by MIND. They did some research with people who have lived or are living in poverty. They asked them about their experiences with mental health support. As part of this research, over 500 people experiencing poverty took part in a survey. More information here: www.mind.org.uk/about-us/our-strategy/working-harder-for-people-facing-poverty/facts-and-figures-about-poverty-and-mental-health/.
2. Check out Sue's lab website, which contains information about the Microbes and Social Equity working group: https://sueishaqlab.org/.
3. Ishaq et al. 2019.
4. Robinson et al., 2022. The Microbes and Social Equity working group was established and several papers have been published including: 'Twenty important research questions in microbial exposure and social equity'.

5. Rauber et al., 2021.
6. Robinson and Jorgensen, 2020.
7. Mears et al., 2019. A relevant study called 'Understanding the socio-economic equity of publicly accessible greenspace distribution: the example of Sheffield, UK'.
8. Walsh et al., 2010. I first learnt of this during a medical sociology lecture at Sheffield. Some areas in Glasgow and other UK cities have similar life expectancies to war-torn countries.
9. The article was called 'No city for old men': www.economist.com/britain/2012/08/25/no-city-for-old-men/.
10. www.glasgowlive.co.uk/news/average-healthy-life-expectancy-glasgow-19693673.
11. www.heraldscotland.com/news/19493561.glasgow-centre-population-health-study-shows-life-expectancy-gap-widening/.
12. Logan, 2015. This is a cracking paper by Alan on social equity.
13. In particular there's an excellent microbiome book, written jointly by Susan and Alan, called *The Secret Life of Your Microbiome*.
14. Liddicoat et al., 2019. Check out some of Craig's other work – his papers are always a pleasure to read.
15. McGurn et al., 2021. This study explored 'Socioeconomic status and determinants of pediatric antibiotic use in the United States'.
16. Parajuli et al., 2018. This is another great paper by the Finnish research group.
17. Kembel et al., 2012.
18. Robinson and Breed, 2019. Part of my PhD involved nature-based interventions. This paper was published with Martin Breed and it describes the many co-benefits of green prescriptions.
19. www.straitstimes.com/singapore/scientists-say-we-breathe-in-1-million-micro-organisms-including-invisible-fungi-every-day.
20. Woo and Lee, 2020.
21. Ishaq et al. 2021.
22. Tessum et al., 2019. They studied how inequity in consumption of goods and services adds to racial-ethnic disparities in air pollution exposure. They created a 'pollution inequity' metric that is generalisable to different pollution types and provides a simple way of 'expressing a disparity between the pollution that people cause and the pollution to which they are exposed'.

Chapter 5: The Psychobiotic Revolution

1. Aung et al., 2013. Traditional Chinese medicine's patterns of disharmony and the Western concept of anxiety disorder are strongly linked.

2. Cryan et al., 2019. Although this attribution is questionable, the wisdom still stands.
3. Escobar, 2017.
4. Robert Whytt was made the first Physician to the King in Scotland in 1761.
5. Miller, 2018.
6. John Abernethy's *Abernethian Code*: http://resource.nlm.nih.gov/61360010R.
7. Baldwin, 1886. Many non-pathogenic microbes floating in the air are now thought to be beneficial to humans.
8. Metchnikoff also published a book called *The Prolongation of Life* in 1907, detailing the gut–health connection.
9. Chu et al., 2019.
10. Ross et al., 2017.
11. Song et al., 2013.
12. Kundu et al., 2017.
13. Cryan et al., 2019.
14. McCoy and Tan, 2014. Find out more about Otto Loewi via this reference.
15. Sun et al., 2019.
16. Fülling et al., 2019. A microbiota–gut–brain axis paper with a great title!
17. Microglia were first identified by Spanish neuroanatomist Pío del Río-Hortega.
18. Abdel-Haq et al., 2019.
19. Van de Wouw et al., 2019. Monocytes can quickly respond to microbial stimuli to inhibit pathogenic microbes at early stages of infection.
20. Brachman et al., 2015. 'Lymphocytes from chronically stressed mice confer antidepressant-like effects'.
21. Aho et al., 2021.
22. The levels of short-chain fatty acids are typically reduced in patients with inflammatory bowel disease.
23. Van de Wouw et al., 2018. Short chain fatty acids can also reduce stress-induced gut–brain axis issues.
24. Check out his other books, too – particularly *Behave* and *A Primate's Memoir.*
25. Smith et al., 2014. And probiotics normalise the gut–brain–microbiota axis in immunodeficient mice.
26. Boziki et al., 2020. Scientists have recently found that infection with Epstein-Barr virus (an invisible foe) dramatically increases the odds of developing MS.

27. Ridaura et al., 2013. The mouse transplant study showing gut microbiota from twins modulate metabolism in mice.

Chapter 6: The Lovebug Effect

1. Robinson and Breed, 2020.
2. Wilson, 1984. Check out Wilson's other books too, including *The Diversity of Life* – this inspired me immensely.
3. Robinson et al., 2021c. Vegetation complexity increases bacterial diversity! This was in an Australia setting. More research is needed to see if these results are representative across biomes.
4. Li et al., 2021.
5. Selway et al., 2020.
6. Liddicoat et al., 2020.
7. Dawkins, 2016. This is a reference to a more recent version of Dawkins's original hypothesis.
8. Alex Honnold is an incredible climber! Check out his documentary movie *Free Solo* – not for those who are afraid of heights.
9. Christensen et al., 2021. Bacteria manipulating flowers!
10. Logan, 2015.

Chapter 7: The Holobiont Blindspot

1. Check out the movie about Lynn called *Symbiotic Earth: How Lynn Margulis Rocked the Boat and Started a Scientific Revolution.*
2. Barrett, 2011. *Beyond the Brain* is a fantastic and thought-provoking book. There is a similar concept called 4^E cognition that is worth exploring.
3. Daniel Kahneman changed 'the way we think about thinking!'. www.theguardian.com/science/2014/feb/16/daniel-kahneman-thinking-fast-and-slow-tributes.
4. Robinson and Cameron, 2020.
5. Bueno-Guerra, 2018.
6. O'Donnell et al., 2020.
7. Westfall et al., 2018. Bacteria can manipulate host feeding decisions in flies called *Drosophila melanogaster.* These are model organisms used in many studies. They share 75% of the genes that cause disease in humans.
8. Sheldrake, 2020, p. 18. Sheldrake is a mycologist and award-winning author.
9. Sharon et al., 2010.
10. Wang et al., 2020.

11. Sapolsky, 2017.
12. Zeki et al., 2004. The authors draw a link between the frontal cortex of the brain and the criminal justice system.

Chapter 8: The Glue that Holds Our Ecosystems Together

1. Innerebner et al., 2011. *Sphingomonas* staves off leaf pathogens in the commonly studied plant *Arabidopsis*.
2. Russell and Ashman, 2019.
3. See Christensen et al., 2021.
4. Falkowski, 2012. Phytoplankton are thought to play a key role in regulating the Earth's climate.
5. See above.
6. Cavicchioli et al., 2019. This is a great paper sending a warning out to society about the threats of climate change and the impacts of microbes.
7. For more information on bioprecipitation, check out work by David Sands from Montana State University.
8. *Caulabacter* in the news! www.nbcnews.com/id/wbna12389963.
9. Penesyan et al., 2021. Current estimates suggests that up to 80% of bacterial and archaeal cells reside in biofilms!

Chapter 9: Microbes and Trees

1. Engelbrecht, 1972.
2. *Wolbachia* is one of the most common parasites on Earth.
3. You can find Suzanne Simard's work here: https://suzannesimard.com/. I also highly recommend her book *Finding the Mother Tree*.
4. Some pollution can change the fungi that provide mineral nutrients to tree roots, causing malnutrition: www.imperial.ac.uk/news/186573/pollution-hits-fungi-that-nourish-european/.
5. Mallet and Roy, 2014. We should view trees as holobionts too.
6. The Incas also created sophisticated canals to bring water to crops.
7. Find out more about his work here: https://standrewsbotanic.org/.
8. Darwin, 1859.
9. Darwin sent many letters to Joseph Hooker. Find out more here: www.darwinproject.ac.uk/.
10. See list of different species concepts here: https://academic.oup.com/sysbio/article/56/6/879/1653163.
11. As above.
12. Check out this article on elm trees in the UK: www.bbc.co.uk/news/science-environment-50519036.

13. More on tree pests and diseases here: www.woodlandtrust.org.uk/trees-woods-and-wildlife/tree-pests-and-diseases.
14. Ulrich et al., 2020.
15. Koskella et al., 2017.
16. Obersteiner et al., 2016. The affected microbiome likely alters the gene expression in the pollen. Reducing pollution could therefore reduce allergies (and the same pollutants can directly affect the human immune system)!

Chapter 10: Rewild, Regenerate, Restore

1. https://en.wikipedia.org/wiki/Homestead_Acts.
2. https://en.wikipedia.org/wiki/Charles_Dana_Wilber.
3. The history of the 'Dust Bowl': www.history.com/topics/great-depression/dust-bowl.
4. President Franklin D. Roosevelt's famous quote was written in 1937. He wrote to all state governors in the US making the case for effective soil management.
5. Professor Duncan Cameron's profile: www.sheffield.ac.uk/biosciences/people/academic-staff/duncan-cameron.
6. Microbes are the foundation of all our farm products: www.youtube.com/watch?v=v5kBu71F2e0.
7. Article on changing levels of nutrients in foods over time: www.scientificamerican.com/article/soil-depletion-and-nutrition-loss/.
8. Davis et al., 2004. I was unable to verify the 26 apples report, but the study in the citation shows dramatic declines in apple nutrients between 1950 and 1999.
9. Fanfield Farm in East Sussex website: https://fanfield.farm/
10. Knepp website: https://knepp.co.uk/home. Also, check out Isabella Tree's book *Wilding*.
11. The '100 harvests' is often spoken about, but there is little scientific evidence to support this claim. However, the fact that we're even talking about this highlights an important issue – soil degradation is happening and it poses a considerable threat to ecosystems and food security. Luckily, people like Chris and Emily at Fanfield Farm are spearheading the solutions.
12. An example of this dynamic between countries can be found in British industrial development. This was predicated on the resources taken from distant 'ghost acres' of land inside the British Empire.
13. Liddicoat et al., 2019.
14. Liddicoat et al., 2020.
15. Robinson et al. 2021b.
16. Wall Kimmerer, 2013.

Chapter 11: Bio-Integrated Design

1. In some countries such as the UK and Australia, over 80% of the population are already urbanised.
2. www.verisk.com/en-gb/3d-visual-intelligence/blog/how-many-tall-buildings-are-there-in-london/.
3. www.youtube.com/watch?v=jxsVG4qGUkE&t=1437s
4. See the abovementioned lecture by Marcos Cruz in note 3.
5. Researchers at the University of Cambridge also created a moss table: www.cam.ac.uk/research/news/the-hidden-power-of-moss.
6. https://news.stanford.edu/news/2010/april/electric-current-plants-041310.html.
7. www.internationale-bauausstellung-hamburg.de/story/iba-hamburg.html.
8. https://english.cw.com.tw/article/article.action?id=1936.
9. Bonnet et al., 2020. Pasteur's work influenced many fields including disease ecology and food production.
10. It turns out the relatively unknown Fanny Hesse is an unsung heroine of science.
11. Research profile: www.ucl.ac.uk/bartlett/architecture/people/mr-richard-beckett.
12. Zhang et al., 2021.
13. The same static electricity occurred when rubbing glass with silk.
14. This was known as the Baghdad Battery: https://en.wikipedia.org/wiki/Baghdad_Battery.
15. www.micro.umass.edu/faculty-and-research/derek-lovley.
16. *Geobacter* protein nanowires: https://en.wikipedia.org/wiki/Bacterial_nanowires.
17. www.cbc.ca/news/science/global-ewaste-monitor-2020-1.5634759.
18. Fungi building in New York City: https://phys.org/news/2021-01-future-homes-fungus.html
19. See link in note 17 above.

Chapter 12: Microbiome-Inspired Green Infrastructure (MIGI)

1. Robinson et al., 2018. My first paper about humans as walking ecosystems.
2. See above. More information in the following blog: www.jakemrobinson.com/blog.
3. Watkins et al., 2020. We are now developing a practical tool so urban managers can optimise their built environments via the microbiome

lens. One exciting application of this is in children's forest schools. Watch this space!

4. Hui et al., 2019.
5. Albrecht, G.A., 2019. *Earth emotions: New words for a new world.* Cornell University Press. A good book that sets the context for the term 'symbiosis'.

Chapter 13: To Catch a Thief: Forensic Microbiology

1. There is some speculation that this was murder. Some evidence suggests that the HIV transmissions were intended by the dentist – for political reasons (see Horowitz, 1994).
2. Robinson et al., 2021d.
3. See above paper.
4. https://innocenceproject.org/.
5. Information at the above link.
6. As above.
7. Sanachai et al., 2016.
8. Carter et al., 2015.
9. Adserias-Garriga et al., 2017.
10. Walker and Datta, 2019.
11. Zeng et al., 2017.
12. Franzosa et al., 2015. Franzosa works at Harvard's Huttenhower Lab and he co-runs an amazing metagenomics course online, which I highly recommend!
13. Meadow et al., 2014. Meadow is a prolific researcher in the microbial forensic space.
14. Kodama et al., 2019. When you die, your body's microbial communities change dramatically over the following hours, days and weeks as the body's tissues decompose.
15. Zhang et al., 2019.
16. Marella et al., 2019.
17. Lee et al., 2017.
18. Robinson et al., 2021d. Our forensic microbiome study provides a review of all the different ways in which microbes are – or could be (in the future) – used to help solve a crime.
19. Zohar Pasternak's profile: www.linkedin.com/in/zohar-pasternak-phd-66b8496b/.
20. More on the Phantom of Heilbronn: https://en.wikipedia.org/wiki/Phantom_of_Heilbronn.

Chapter 14: Microbes in Outer Space

1. Fred Hoyle and Chandra Wickramasinghe were influential proponents of panspermia.
2. *Deinococcus radiodurans* was discovered by Arthur Anderson in Oregon in 1956.
3. Kawaguchi et al., 2020.
4. www.genomenewsnetwork.org/articles/07_02/deinococcus.shtml.
5. See Kawaguchi et al., 2020.
6. Smith, 2018.
7. *C. vulgaris* is also used in many other fields, from foods to waste-water treatment.
8. https://en.wikipedia.org/wiki/Johann_August_Ephraim_Goeze.
9. www.sciencefocus.com/nature/amazing-facts-about-tardigrades-the-worlds-toughest-animal/.
10. https://solarsystem.nasa.gov/missions/beresheet/in-depth/.
11. Traspas and Burchell, 2021.
12. Check out this video: www.youtube.com/watch?v=thCuPSn7Hq4.

Chapter 15: You Are What Your Microbes Eat

1. https://imagindairy.com/.
2. Dairy cow gases: www.ucdavis.edu/food/news/making-cattle-more-sustainable.
3. European consumers are including more plant-based foods in their diets: https://smartproteinproject.eu/wp-content/uploads/FINAL_Pan-EU-consumer-survey_Overall-Report-.pdf.
4. https://ourworldindata.org/agricultural-land-by-global-diets.
5. Raymond Lindeman first proposed the '10% rule' in 1942: https://study.com/academy/lesson/the-10-energy-rule-in-a-food-chain.html.
6. Button and Dutton, 2012.
7. www.youtube.com/watch?v=axrywDP9Ii0.
8. www.smithsonianmag.com/smart-news/bread-was-made-using-4500-year-old-egyptian-yeast-180972842/.
9. www.sciencedaily.com/releases/2007/12/071205140118.htm.
10. Pascoe, 2014. *Dark Emu* re-examines colonial accounts of Aboriginal Peoples in Australia.
11. Madden et al., 2018.
12. Stefanini et al., 2012. Yeasts are common gut residents of many other insects, such as *Drosophila* fruit flies and mosquitoes.
13. Meriggi et al., 2019.

14. Carrigan et al., 2015. Early hominids are thought to have adapted to metabolise ethanol long before humans purposefully fermented foods.
15. www.homegrounds.co/uk/history-of-coffee/.
16. www.nytimes.com/2021/08/13/well/eat/yogurt-kimchi-kombucha-microbiome.html.
17. Kim and Adhikari, 2020.
18. https://kingfm.com/129-year-old-sourdough-starter-is-still-growing-in-newcastle-wyoming/.
19. www.mercurynews.com/2015/11/23/san-franciscos-boudin-bakery-serves-up-a-taste-of-history-in-each-bite/.
20. Conlon and Bird, 2015. Saturated fats may increase pro-inflammatory gut microbes by stimulating the formation of bile acids.
21. Tomova et al., 2019. Vegan diets (comprising of mostly wholefoods) are typically associated with lower body weight and likely benefit microbial diversity and protect against inflammation.
22. South Korea has also seen reductions in infant mortality and in infections and blood pressure conditions due to lifestyle changes and advances in medical practices. Microbiome-nourishing food such as kimchi may play an important role in improving life expectancy: www.weforum.org/agenda/2017/07/south-korean-women-life-expectancy-kimchi/.
23. https://lightorangebean.com/.

Chapter 16: Nature Connectedness

1. Wilson, 1992. *The Diversity of Life* – one of my inspirations for learning about ecology.
2. Wilson, 1992, p. 7.
3. Thomas et al., 2021.
4. Survey commissioned by the app Hoop. More here: https://www.earlham.ac.uk/articles/shocking-state-biodiversity-education-and-mass-extinction.
5. Professor Miles Richardson: www.derby.ac.uk/staff/miles-richardson/.
6. See Wall Kimmerer, 2013 for more.
7. Robinson et al., 2021a.
8. Pascoe, 2014.
9. Gammage, 2013. This book contains a similar narrative to *Dark Emu* but is more academically oriented.
10. www.biblegateway.com/passage/?search=Genesis%201&version=ESV.
11. www.vatican.va/content/francesco/en/encyclicals/documents/papa-francesco_20150524_enciclica-laudato-si.html.

12. https://nature.berkeley.edu/departments/espm/env-hist/Moses.pdf.
13. Bacon and Montagu, 1857.
14. De Vos et al., 2015. Species extinction estimations.
15. Leopold, 1949, p. 320. *A Sand County Almanac* – another of my favourite books.
16. Leopold, 1949, p. 320.
17. Barrera-Hernández et al., 2020. Nature connectedness predicts environmental stewardship.
18. Walker, 2017.
19. Nisbet et al., 2009.
20. Richardson et al., 2019.
21. Nisbet and Zelenski, 2013. This is a super-quick and easy test of nature connectedness.
22. Robinson et al. 2021e.

Appendix: Microbes 101

1. To find out more about Carl Woese, I recommend reading *The Tangled Tree* by David Quammen.
2. Morell, 1996.
3. Dance, 2021. A fascinating article about viruses!
4. See above reference.
5. Loron et al., 2019.
6. Falkowski, 2012. Phytoplankton include drifting plants, algae and some bacteria that can photosynthesise. Species in the *Prochlorococcus* genus are the smallest photosynthetic organisms on the planet.
7. Fields et al., 2020. Consuming *Chlamydomonas reinhardtii* didn't cause any changes in the study subjects' gut microbiome.

Glossary

1. Haahtela, 2019.
2. European Commission's Green Infrastructure Strategy, 2013.
3. Bordenstein and Theis, 2015.
4. Strachan et al., 1989.
5. Kahneman, 2011.

Acknowledgements

With enormous gratitude to my wife, Kate who has supported me throughout my book-writing journey. Thank you also to my family and friends whose enthusiasm helped me reach the finish line, even if I fail to make sense to you half of the time! Anyone who has been a part of my life, past and present, you have influenced my worldview. Our shared experiences have contributed to the stories in this book. Thank you. Many scientists from different corners of world informed the academic contents of the book. You injected life into the chapters, and for this, I thank you dearly. The book wouldn't be a book without you. In order of appearance, thanks to Prof. Graham Rook, Dr Suzanne Ishaq, Prof. Susan Prescott, Dr Alan Logan, Prof. John F. Cryan, Dr Gillian Orrow, Prof. Robert Sapolsky, Prof. Suzanne Simard, Chris and Emily Huskins, Dr Martin F. Breed, Dr Craig Liddicoat, Chris Cando-Dumancela, Dr Harry Watkins, Dr Chris Skelly, Prof. Robin Wall-Kimmerer, Prof. Daniel Kahnemann, Prof. Louise Barrett, Dr Brenda Parker, Prof. Marcos Cruz, William Scott, Prof. Anna Jorgensen, Dr Ross Cameron, Dr Paul Brindley, Prof. Eran Elhaik, Prof. Miles Richardson, and the late Prof. E.O. Wilson and Prof. Lynn Margulis. A huge thanks to the managers, production and marketing staff at Pelagic Publishing (Nigel, David, Sarah) and to Sara the copy editor; you have improved the readability and reach of the book tremendously. A big thanks to Laura Brett who did the beautiful cover art. Thank you to those who read the book and provided feedback – my mother (Maureen), father (Peter) and sister (Hannah), my mother-in-law (Jean), Joe Hodges, Dr David Fuller, Vicky Peace, Rachel White, Dr Martin Breed (again), and Esther Breed. Thanks also to nature authors Dr Rebecca Nesbit and Lucy Jones who provided guidance when I first tried to navigate the publishing jungle. And finally, thanks to all the invisible friends that keep us alive, that shape the ecosystems we live in, the wildlife we love, the foods we eat, the air we breathe.

References

Abdel-Haq, R., Schlachetzki, J.C., Glass, C.K. and Mazmanian, S.K., 2019. Microbiome–microglia connections via the gut–brain axis. *Journal of Experimental Medicine*, 216(1), pp. 41–59.

Adserias-Garriga, J., Hernández, M., Quijada, N.M., Lázaro, D.R., Steadman, D. and Garcia-Gil, J., 2017. Daily thanatomicrobiome changes in soil as an approach of postmortem interval estimation: an ecological perspective. *Forensic Science International*, 278, pp. 388–395.

Aho, V.T., Houser, M.C., Pereira, P.A., Chang, J., Rudi, K., Paulin, L., Hertzberg, V., Auvinen, P., Tansey, M.G. and Scheperjans, F., 2021. Relationships of gut microbiota, short-chain fatty acids, inflammation, and the gut barrier in Parkinson's disease. *Molecular Neurodegeneration*, 16(1), pp. 1–14.

Aminov, R.I., 2010. A brief history of the antibiotic era: lessons learned and challenges for the future. *Frontiers in Microbiology*, 1, p. 134.

Atchison, R.W., Casto, B.C. and Hammon, W.M., 1965. Adenovirus-associated defective virus particles. *Science*, 149(3685), pp. 754–755.

Aung, S.K., Fay, H. and Hobbs, R.F., 2013. Traditional Chinese medicine as a basis for treating psychiatric disorders: a review of theory with illustrative cases. *Medical Acupuncture*, 25(6), pp. 398–406.

Bacon, F. and Montagu, B., 1857. *The Works of Francis Bacon (Vol. 4)*. Parry & McMillan.

Baldwin, B.J., 1886. Is there air without germs? *Medical Record* (1866-1922), 30(15), p. 417.

Barrera-Hernández, L.F., Sotelo-Castillo, M.A., Echeverría-Castro, S.B. and Tapia-Fonllem, C.O., 2020. Connectedness to nature: its impact on sustainable behaviors and happiness in children. *Frontiers in Psychology*, p. 276.

Barrett, L., 2011. *Beyond the Brain*. Princeton University Press.

Bear, I.J. and Thomas, R.G., 1964. Nature of argillaceous odour. *Nature*, 201(4923), pp. 993–995.

Becher, P.G., Verschut, V., Bibb, M.J., Bush, M.J., Molnár, B.P., Barane, E., Al-Bassam, M.M., Chandra, G., Song, L., Challis, G.L. and Buttner, M.J., 2020. Developmentally regulated volatiles geosmin and 2-methylisoborneol attract a soil arthropod to *Streptomyces* bacteria promoting spore dispersal. *Nature Microbiology*, 5(6), pp. 821–829.

Berg, G., Rybakova, D., Fischer, D., Cernava, T., Vergès, M.C.C., Charles, T., Chen, X., Cocolin, L., Eversole, K., Corral, G.H. and Kazou, M., 2020. Microbiome definition re-visited: old concepts and new challenges. *Microbiome*, 8(1), pp. 1–22.

Bolourian, A. and Mojtahedi, Z., 2018. Streptomyces, shared microbiome member of soil and gut, as 'old friends' against colon cancer. *FEMS Microbiology Ecology*, 94(8), p.fiy120.

Bonnet, M., Lagier, J.C., Raoult, D. and Khelaifia, S., 2020. Bacterial culture through selective and non-selective conditions: the evolution of culture media in clinical microbiology. *New Microbes and New Infections*, 34, p. 100622.

Bordenstein, S.R. and Theis, K.R., 2015. Host biology in light of the microbiome: ten principles of holobionts and hologenomes. *PLoS Biology*, 13(8), p. e1002226.

Boziki, M.K., Kesidou, E., Theotokis, P., Mentis, A.F.A., Karafoulidou, E., Melnikov, M., Sviridova, A., Rogovski, V., Boyko, A. and Grigoriadis, N., 2020. Microbiome in multiple sclerosis: Where are we, what we know and do not know. *Brain Sciences*, 10(4), p. 234.

Brachman, R.A., Lehmann, M.L., Maric, D. and Herkenham, M., 2015. Lymphocytes from chronically stressed mice confer antidepressant-like effects to naive mice. *Journal of Neuroscience*, 35(4), pp. 1530–1538.

Breitbart, M., Bonnain, C., Malki, K. and Sawaya, N.A., 2018. Phage puppet masters of the marine microbial realm. *Nature Microbiology*, 3(7), pp. 754–766.

Bueno-Guerra, N., 2018. How to apply the concept of Umwelt in the evolutionary study of cognition. *Frontiers in Psychology*, p. 2001.

Button, J.E. and Dutton, R.J., 2012. Cheese microbes. *Current Biology*, 22(15), pp. R587-R589.

Carrigan, M.A., Uryasev, O., Frye, C.B., Eckman, B.L., Myers, C.R., Hurley, T.D. and Benner, S.A., 2015. Hominids adapted to metabolize ethanol long before human-directed fermentation. *Proceedings of the National Academy of Sciences*, 112(2), pp. 458–463.

Carter, D.O., Metcalf, J.L., Bibat, A. and Knight, R., 2015. Seasonal variation of postmortem microbial communities. *Forensic Science, Medicine, and Pathology*, 11(2), pp. 202–207.

Cavicchioli, R., Ripple, W.J., Timmis, K.N., Azam, F., Bakken, L.R., Baylis, M., Behrenfeld, M.J., Boetius, A., Boyd, P.W., Classen, A.T. and Crowther, T.W., 2019. Scientists' warning to humanity: microorganisms and climate change. *Nature Reviews Microbiology*, 17(9), pp. 569–586.

Chen, Q.L., An, X.L., Zheng, B.X., Gillings, M., Peñuelas, J., Cui, L., Su, J.Q. and Zhu, Y.G., 2019. Loss of soil microbial diversity exacerbates spread of antibiotic resistance. *Soil Ecology Letters*, 1(1), pp. 3–13.

Christensen, S.M., Munkres, I. and Vannette, R.L., 2021. Nectar bacteria stimulate pollen germination and bursting to enhance microbial fitness. *Current Biology*, 31(19), pp. 4373–4380.

Chu, C., Murdock, M.H., Jing, D., Won, T.H., Chung, H., Kressel, A.M., Tsaava, T., Addorisio, M.E., Putzel, G.G., Zhou, L. and Bessman, N.J., 2019. The microbiota regulate neuronal function and fear extinction learning. *Nature*, 574(7779), pp. 543–548.

Conlon, M.A. and Bird, A.R., 2015. The impact of diet and lifestyle on gut microbiota and human health. *Nutrients*, 7(1), pp. 17–44.

Cryan, J.F., O'Riordan, K.J., Cowan, C.S., Sandhu, K.V., Bastiaanssen, T.F., Boehme, M., Codagnone, M.G., Cussotto, S., Fulling, C., Golubeva, A.V. and Guzzetta, K.E., 2019. The microbiota-gut-brain axis. *Physiological Reviews*, 99 (4), pp. 1877–2013.

Dance, A., 2021. Beyond coronavirus: the virus discoveries transforming biology. *Nature*, 595(7865), pp. 22–25.

Darwin, C., 1859. *On the origin of species by means of natural selection or the preservation of favoured races in the struggle for life*, Vol. 2. London: John Murray.

Davis, D.M., 2020. *The Beautiful Cure*. Chicago: University of Chicago Press.

Davis, D.R., Epp, M.D. and Riordan, H.D., 2004. Changes in USDA food composition data for 43 garden crops, 1950 to 1999. *Journal of the American College of Nutrition*, 23(6), pp. 669–682.

Dawkins, R., 2016. *The Extended Phenotype: The Long Reach of the Gene*. Oxford: Oxford University Press.

De Vos, J.M., Joppa, L.N., Gittleman, J.L., Stephens, P.R. and Pimm, S.L., 2015. Estimating the normal background rate of species extinction. *Conservation Biology*, 29(2), pp. 452–462.

Engelbrecht, L., 1972. Cytokinins in leaf-cuttings of *Phaseolus vulgaris* L. during their development. *Biochemie und Physiologie der Pflanzen*, 163(4), pp. 335–343.

Escobar, A., 2017. Sustaining the pluriverse: the political ontology of territorial struggles in Latin America. In M. Brightman and J. Lewis (eds) *The Anthropology of Sustainability: Beyond Development and Progress*, pp. 237–256. Palgrave Studies in the Anthropology of Sustainability. New York: Palgrave.

European Commission's Green Infrastructure Strategy, 2013. Available at: https://ec.europa.eu/environment/nature/ecosystems/strategy/index_en.htm#:~:text=On%206%20May%202013%2C%20the,its%20many%20benefits%20to%20us.

Falkowski, P., 2012. Ocean science: the power of plankton. *Nature*, 483(7387), pp. S17–S20.

Fields, F.J., Lejzerowicz, F., Schroeder, D., Ngoi, S.M., Tran, M., McDonald, D., Jiang, L., Chang, J.T., Knight, R. and Mayfield, S., 2020. Effects of the microalgae *Chlamydomonas* on gastrointestinal health. *Journal of Functional Foods*, 65, p. 103738.

Fierer, N., Hamady, M., Lauber, C.L. and Knight, R., 2008. The influence of sex, handedness, and washing on the diversity of hand surface bacteria. *Proceedings of the National Academy of Sciences*, 105(46), pp. 17994–17999.

Flanders, S.A. and Saint, S., 2014. Why does antimicrobial overuse in hospitalized patients persist? *JAMA Internal Medicine*, 174(5), pp. 661–662.

Franzosa, E.A., Huang, K., Meadow, J.F., Gevers, D., Lemon, K.P., Bohannan, B.J. and Huttenhower, C., 2015. Identifying personal microbiomes using metagenomic codes. *Proceedings of the National Academy of Sciences*, 112(22), pp. E2930–E2938.

Fülling, C., Dinan, T.G. and Cryan, J.F., 2019. Gut microbe to brain signaling: what happens in vagus .... *Neuron*, 101(6), pp. 998–1002.

Gammage, B., 2013. *The Biggest Estate on Earth: How Aborigines Made Australia*. Sydney: Allen and Unwin.

Gilbert, J.A. and Neufeld, J.D., 2014. Life in a world without microbes. *PLoS Biology*, 12(12), p. e1002020.

Haahtela, T., Laatikainen, T., Alenius, H., Auvinen, P., Fyhrquist, N., Hanski, I., Von Hertzen, L., Jousilahti, P., Kosunen, T.U., Markelova, O. and Mäkelä, M.J., 2015. Hunt for the origin of allergy – comparing the Finnish and Russian Karelia. *Clinical & Experimental Allergy*, 45(5), pp. 891–901.

Haahtela, T., 2019. A biodiversity hypothesis. *Allergy*, 74(8), pp. 1445–1456.

Hesselmar, B., Hicke-Roberts, A., Lundell, A.C., Adlerberth, I., Rudin, A., Saalman, R., Wennergren, G. and Wold, A.E., 2018. Pet-keeping in early life reduces the risk of allergy in a dose-dependent fashion. *PLOS One*, 13(12), p. e0208472.

Honeker, L.K., Sharma, A., Gozdz, J., Theriault, B., Patil, K., Gimenes, Jr, J.A., Horner, A., Pivniouk, V., Igartua, C., Stein, M.M. and Holbreich, M., 2019. Gut microbiota from Amish but not Hutterite children protect germ-free mice from experimental asthma. In *D92. The Microbiome and Lung Disease*, pp. A7022–A7022. American Thoracic Society.

Horowitz, L.G., 1994. Murder and cover-up could explain the Florida dental AIDS mystery. *British Dental Journal*, 177(11), pp. 423–427.

Hui, N., Grönroos, M., Roslund, M.I., Parajuli, A., Vari, H.K., Soininen, L., Laitinen, O.H., Sinkkonen, A. and ADELE Research Group, 2019. Diverse environmental microbiota as a tool to augment biodiversity in urban landscaping materials. *Frontiers in Microbiology*, 10, p. 536.

Innerebner, G., Knief, C. and Vorholt, J.A., 2011. Protection of *Arabidopsis thaliana* against leaf-pathogenic *Pseudomonas syringae* by *Sphingomonas* strains in a controlled model system. *Applied and Environmental Microbiology*, 77(10), pp. 3202–3210.

Ishaq, S.L., Rapp, M., Byerly, R., McClellan, L.S., O'Boyle, M.R., Nykanen, A., Fuller, P.J., Aas, C., Stone, J.M., Killpatrick, S. and Uptegrove, M.M., 2019. Framing the discussion of microorganisms as a facet of social equity in human health. *PLOS Biology*, 17(11), p. e3000536.

Ishaq, S.L., Hotopp, A., Silverbrand, S., Dumont, J.E., Michaud, A., MacRae, J.D., Stock, S.P. and Groden, E., 2021. Bacterial transfer from Pristionchus entomophagus nematodes to the invasive ant Myrmica rubra and the potential for colony mortality in coastal Maine. *Iscience*, 24(6), p.102663.

John Innes Centre. 2020. Geosmin. Available at: https://phys.org/news/2020-04-unearths-science.html

Kahneman, D., 2011. *Thinking, fast and slow*. London: Macmillan.

Kannadan, A., 2018. History of the miasma theory of disease. *ESSAI*, 16(1), p. 18.

Kawaguchi, Y., Shibuya, M., Kinoshita, I., Yatabe, J., Narumi, I., Shibata, H., Hayashi, R., Fujiwara, D., Murano, Y., Hashimoto, H. and Imai, E., 2020. DNA damage and survival time course of deinococcal cell pellets during 3 years of exposure to outer space. *Frontiers in Microbiology*, 11, p. 2050.

Kembel, S.W., Jones, E., Kline, J., Northcutt, D., Stenson, J., Womack, A.M., Bohannan, B.J., Brown, G.Z. and Green, J.L., 2012. Architectural design influences the diversity and structure of the built environment microbiome. *The ISME Journal*, 6(8), pp. 1469–1479.

Kim, J. and Adhikari, K., 2020. Current trends in kombucha: marketing perspectives and the need for improved sensory research. *Beverages*, 6(1), p. 15.

Kirchhelle, C., 2018. Pharming animals: a global history of antibiotics in food production (1935–2017). *Palgrave Communications*, 4(1), pp. 1–13.

Kodama, W.A., Xu, Z., Metcalf, J.L., Song, S.J., Harrison, N., Knight, R., Carter, D.O. and Happy, C.B., 2019. Trace evidence potential in postmortem skin microbiomes: from death scene to morgue. *Journal of Forensic Sciences*, 64(3), pp. 791–798.

Koskella, B., Meaden, S., Crowther, W.J., Leimu, R. and Metcalf, C.J.E., 2017. A signature of tree health? Shifts in the microbiome and the ecological drivers of horse chestnut bleeding canker disease. *New Phytologist*, 215(2), pp. 737–746.

Kundu, P., Blacher, E., Elinav, E. and Pettersson, S., 2017. Our gut microbiome: the evolving inner self. *Cell*, 171(7), pp. 1481–1493.

Larsen, J., Raisen, C.L., Ba, X. et al., 2022. Emergence of methicillin resistance predates the clinical use of antibiotics. *Nature*. Epub ahead of print. PMID: 34987223.

Lee, S.Y., Woo, S.K., Lee, S.M., Ha, E.J., Lim, K.H., Choi, K.H., Roh, Y.H. and Eom, Y.B., 2017. Microbiota composition and pulmonary surfactant protein expression as markers of death by drowning. *Journal of Forensic Sciences*, 62(4), pp. 1080–1088.

Leo, E.E.M. and Campos, M.R.S., 2020. Effect of ultra-processed diet on gut microbiota and thus its role in neurodegenerative diseases. *Nutrition*, 71, p. 110609.

Leopold, A., 1949. *A Sand County Almanac (Outdoor Essays & Reflections)*. New York: Ballantine (reprint 1966 and 1990).

Li, H., Wu, Z.F., Yang, X.R., An, X.L., Ren, Y. and Su, J.Q., 2021. Urban greenness and plant species are key factors in shaping air microbiomes and reducing airborne pathogens. *Environment International*, 153, p. 106539.

Li, S., Yao, Q., Liu, J., Yu, Z., Li, Y., Jin, J., Liu, X. and Wang, G., 2022. Liming mitigates the spread of antibiotic resistance genes in an acid black soil. *Science of The Total Environment*, 817, p.152971.

Liddicoat, C., Weinstein, P., Bissett, A., Gellie, N.J., Mills, J.G., Waycott, M. and Breed, M.F., 2019. Can bacterial indicators of a grassy woodland restoration inform ecosystem assessment and microbiota-mediated human health? *Environment International*, 129, pp. 105–117.

Liddicoat, C., Sydnor, H., Cando-Dumancela, C., Dresken, R., Liu, J., Gellie, N.J., Mills, J.G., Young, J.M., Weyrich, L.S., Hutchinson, M.R. and Weinstein, P., 2020. Naturally-diverse airborne environmental microbial exposures modulate the gut microbiome and may provide anxiolytic benefits in mice. *Science of the Total Environment*, 701, p. 134684.

Locey, K.J. and Lennon, J.T., 2016. Scaling laws predict global microbial diversity. *Proceedings of the National Academy of Sciences*, 113(21), pp. 5970–5975.

Logan, A.C., 2015. Dysbiotic drift: mental health, environmental grey space, and microbiota. *Journal of Physiological Anthropology*, 34(1), pp. 1–16.

Loron, C.C., Rainbird, R.H., Turner, E.C., Greenman, J.W. and Javaux, E.J., 2019. Organic-walled microfossils from the late Mesoproterozoic to early Neoproterozoic lower Shaler Supergroup (Arctic Canada): diversity and biostratigraphic significance. *Precambrian Research*, 321, pp. 349–374.

Lowry, C.A., Smith, D.G., Siebler, P.H., Schmidt, D., Stamper, C.E., Hassell, J.E., Yamashita, P.S., Fox, J.H., Reber, S.O., Brenner, L.A. and Hoisington, A.J., 2016. The microbiota, immunoregulation, and mental health: implications for public health. *Current Environmental Health Reports*, 3(3), pp. 270–286.

Ma, Z.S. and Mei, J., 2022. Stochastic neutral drifts seem prevalent in driving human virome assembly: Neutral, near-neutral and non-neutral theoretic analyses. *Computational and Structural Biotechnology Journal*, 20, pp. 2029–2041.

Madden, A.A., Epps, M.J., Fukami, T., Irwin, R.E., Sheppard, J., Sorger, D.M. and Dunn, R.R., 2018. The ecology of insect–yeast relationships and its relevance to human industry. *Proceedings of the Royal Society B: Biological Sciences*, 285(1875), p. 20172733.

Mallet, P.L. and Roy, S., 2014. The symbiosis between *Frankia alni* and alder shrubs results in a tolerance of the environmental stress associated with tailings from the Canadian oil sands industry. *Journal of Petroleum and Environmental Biotechnology*, 5(3), pp. 1–9.

Marella, G.L., Feola, A., Marsella, L.T., Mauriello, S., Giugliano, P. and Arcudi, G., 2019. Diagnosis of drowning, an everlasting challenge in forensic medicine: review of the literature and proposal of a diagnostic algorithm. *Acta Med*, 35, pp. 900–919.

McCoy, A.N. and Tan, Y.S., 2014. Otto Loewi (1873–1961): Dreamer and Nobel laureate. *Singapore Medical Journal*, 55(1), p. 3.

McGurn, A., Watchmaker, B., Adam, K., Ni, J., Babinski, P., Friedman, H., Boyd, B., Dugas, L.R. and Markossian, T., 2021. Socioeconomic status and determinants of pediatric antibiotic use. *Clinical Pediatrics*, 60(1), pp. 32–41.

Meadow, J.F., Altrichter, A.E. and Green, J.L., 2014. Mobile phones carry the personal microbiome of their owners. *PeerJ*, 2, p. e447.

Mears, M., Brindley, P., Maheswaran, R. and Jorgensen, A., 2019. Understanding the socioeconomic equity of publicly accessible greenspace distribution: the example of Sheffield, UK. *Geoforum*, 103, pp. 126–137.

Meriggi, N., Di Paola, M., Vitali, F., Rivero, D., Cappa, F., Turillazzi, F., Gori, A., Dapporto, L., Beani, L., Turillazzi, S. and Cavalieri, D., 2019. *Saccharomyces cerevisiae* induces immune enhancing and shapes gut microbiota in social wasps. *Frontiers in Microbiology*, p. 2320.

Microbiology by numbers, 2011. *Nature Reviews Microbiology*, 9, p. 628.

Miller, I., 2018. The gut–brain axis: historical reflections. *Microbial Ecology in Health and isease*, 29(2), p. 1542921.

Morell, V., 1996. Life's last domain: with the genome of the archaeon microbe *Methanococcus jannaschii* sequenced, researchers now have genomes for life's three domains. And only 44% of the archaeon's genes are familiar. *Science*, 273(5278), pp. 1043–1045.

Ngure, F.M., Humphrey, J.H., Mbuya, M.N., Majo, F., Mutasa, K., Govha, M., Mazarura, E., Chasekwa, B., Prendergast, A.J., Curtis, V. and Boor, K.J., 2013. Formative research on hygiene behaviors and geophagy among infants and young children and implications of exposure to fecal bacteria. *The American Journal of Tropical Medicine and Hygiene*, 89(4), p. 709.

Nisbet, E.K. and Zelenski, J.M., 2013. The NR-6: a new brief measure of nature relatedness. *Frontiers in Psychology*, 4, p. 813.

Nisbet, E.K., Zelenski, J.M. and Murphy, S.A., 2009. The nature relatedness scale: Linking individuals' connection with nature to environmental concern and behavior. *Environment and Behavior*, 41(5), pp. 715–740.

Nyanza, E.C., Joseph, M., Premji, S.S., Thomas, D.S. and Mannion, C., 2014. Geophagy practices and the content of chemical elements in the soil eaten by pregnant women in artisanal and small scale gold mining communities in Tanzania. *BMC Pregnancy and Childbirth*, 14(1), pp. 1–10.

Obersteiner, A., Gilles, S., Frank, U., Beck, I., Häring, F., Ernst, D., Rothballer, M., Hartmann, A., Traidl-Hoffmann, C. and Schmid, M., 2016. Pollen-associated microbiome correlates with pollution parameters and the allergenicity of pollen. *PLOS One*, 11(2), p. e0149545.

O'Donnell, M.P., Fox, B.W., Chao, P.H., Schroeder, F.C. and Sengupta, P., 2020. A neurotransmitter produced by gut bacteria modulates host sensory behaviour. *Nature*, 583(7816), pp. 415–420.

Osawa, R., Blanshard, W.H. and Ocallaghan, P.G., 1993. Microbiological studies of the intestinal microflora of the koala, Phascolarctos-cinereus. 2. Pap, a special maternal feces consumed by juvenile koalas. *Australian Journal of oology*, 41(6), pp. 611–620.

Parajuli, A., Grönroos, M., Siter, N., Puhakka, R., Vari, H.K., Roslund, M.I., Jumpponen, A., Nurminen, N., Laitinen, O.H., Hyöty, H. and Rajaniemi, J., 2018. Urbanization reduces transfer of diverse environmental microbiota indoors. *Frontiers in Microbiology*, 9, p. 84.

Pascoe, B., 2014. *Dark Emu: Black Seeds: Agriculture or Accident?* Broome: Magabala Books.

Penesyan, A., Paulsen, I.T., Kjelleberg, S. and Gillings, M.R., 2021. Three faces of biofilms: a microbial lifestyle, a nascent multicellular organism, and an incubator for diversity. *npj Biofilms and Microbiomes*, 7(1), pp. 1–9.

Peterson, E. and Kaur, P., 2018. Antibiotic resistance mechanisms in bacteria: relationships between resistance determinants of antibiotic producers, environmental bacteria, and clinical pathogens. *Frontiers in Microbiology*, 9, p. 2928.

Rahman, M.S., Hassan, M.M. and Chowdhury, S., 2021. Determination of antibiotic residues in milk and assessment of human health risk in Bangladesh. *Heliyon*, 7(8), p. e07739.

Rauber, F., Chang, K., Vamos, E.P., da Costa Louzada, M.L., Monteiro, C.A., Millett, C. and Levy, R.B., 2021. Ultra-processed food consumption and risk of obesity: a prospective cohort study of UK Biobank. *European Journal of Nutrition*, 60(4), pp. 2169–2180.

Richardson, M., Hunt, A., Hinds, J., Bragg, R., Fido, D., Petronzi, D., Barbett, L., Clitherow, T. and White, M., 2019. A measure of nature connectedness for children and adults: validation, performance, and insights. *Sustainability*, 11(12), p. 3250.

Ridaura, V.K., Faith, J.J., Rey, F.E., Cheng, J., Duncan, A.E., Kau, A.L., Griffin, N.W., Lombard, V., Henrissat, B., Bain, J.R. and Muehlbauer, M.J., 2013. Gut microbiota from twins discordant for obesity modulate metabolism in mice. *Science*, 341(6150), p. 1241214.

Robinson, J.M. and Breed, M.F., 2019. Green prescriptions and their co-benefits: integrative strategies for public and environmental health. *Challenges*, 10(1), p. 9.

Robinson, J.M. and Breed, M.F., 2020. The Lovebug Effect: is the human biophilic drive influenced by interactions between the host, the environment, and the microbiome? *Science of the Total Environment*, 720, p. 137626.

Robinson, J.M. and Cameron, R., 2020. The holobiont blindspot: relating host-microbiome interactions to cognitive biases and the concept of the 'Umwelt'. *Frontiers in Psychology*, 11, p. 3255.

Robinson, J.M. and Jorgensen, A., 2020. Rekindling old friendships in new landscapes: the environment–microbiome–health axis in the realms of landscape research. *People and Nature*, 2(2), pp. 339–349.

Robinson, J.M., Mills, J.G. and Breed, M.F., 2018. Walking ecosystems in microbiome-inspired green infrastructure: an ecological perspective on enhancing personal and planetary health. *Challenges*, 9(2), p. 40.

Robinson, J.M., Cameron, R. and Jorgensen, A., 2021a. Germaphobia! Does our relationship with and knowledge of biodiversity affect our attitudes toward microbes? *Frontiers in Psychology*, 12, p. 2520.

Robinson, J.M., Cameron, R. and Parker, B., 2021b. The effects of anthropogenic sound and artificial light exposure on microbiomes: ecological and public health implications. *Frontiers in Ecology and Evolution*, p. 321.

Robinson, J.M., Cando-Dumancela, C., Antwis, R.E., Cameron, R., Liddicoat, C., Poudel, R., Weinstein, P. and Breed, M.F., 2021c. Exposure to airborne bacteria depends upon vertical stratification and vegetation complexity. *Scientific Reports*, 11(1), pp. 1–16.

Robinson, J.M., Pasternak, Z., Mason, C.E. and Elhaik, E., 2021d. Forensic applications of microbiomics: a review. *Frontiers in Microbiology*, 11, p. 3455.

Robinson, J.M., Brindley, P., Cameron, R., MacCarthy, D. and Jorgensen, A., 2021e. Nature's role in supporting health during the COVID-19 pandemic: A geospatial and socioecological study. *International ournal of Environmental Research and Public Health*, 18(5), p. 2227.

Robinson, J.M., Redvers, N., Camargo, A., Bosch, C.A., Breed, M.F., Brenner, L.A., Carney, M.A., Chauhan, A., Dasari, M., Dietz, L.G. and Friedman, M., 2022. Twenty important research questions in microbial exposure and social equity. *mSystems* 7(1), e01240-21.

Rodriguez, B., Prioult, G., Hacini-Rachinel, F., Moine, D., Bruttin, A., Ngom-Bru, C., Labellie, C., Nicolis, I., Berger, B., Mercenier, A. and Butel, M.J., 2012. Infant gut microbiota is protective against cow's milk allergy in mice despite immature ileal T-cell response. *FEMS Microbiology Ecology*, 79(1), pp. 192–202.

Rook, A., 1968. Skin diseases caused by arthropods and other venomous or noxious animals. In A. Rook, D.S. Wilkinson and F.J. Ebling (eds) *Textbook of Dermatology*. pp.1-3683. Malden: Blackwell.

Rook, G.A. and Bloomfield, S.F., 2021. Microbial exposures that establish immunoregulation are compatible with targeted hygiene. *Journal of Allergy and Clinical Immunology*, 148(1), pp. 33–39.

Rook, G.A., Martinelli, R. and Brunet, L.R., 2005. The 'old friends' hypothesis; how early contact with certain microorganisms may influence immunoregulatory circuits. *Perinatal Programming*, pp. 183–194.

Roslund, M.I., Puhakka, R., Grönroos, M., Nurminen, N., Oikarinen, S., Gazali, A.M., Cinek, O., Kramná, L., Siter, N., Vari, H.K. and Soininen, L., 2020. Biodiversity intervention enhances immune regulation and health-associated commensal microbiota among daycare children. *Science Advances*, 6(42), p. eaba2578.

Ross, A.A., Doxey, A.C. and Neufeld, J.D., 2017. The skin microbiome of cohabiting couples. *MSystems*, 2(4), pp. e00043-17.

Russell, A.L. and Ashman, T.L., 2019. Associative learning of flowers by generalist bumble bees can be mediated by microbes on the petals. *Behavioral Ecology*, 30(3), pp. 746–755.

Sagan, L., 1967. On the origin of mitosing cells. *Journal of Theoretical Biology*, 14(3), pp. 225-IN6.

Sanachai, A., Katekeaw, S. and Lomthaisong, K., 2016. Forensic soil investigation from the 16S rDNA profiles of soil bacteria obtained by denaturing gradient gel electrophoresis. *Chiang Mai Journal of Science*, 43(4).

Sapolsky, R.M., 2017. *Behave: The Biology of Humans at Our Best and Worst.* London: Penguin.

Schulz, H.N., 2002. *Thiomargarita namibiensis*: giant microbe holding its breath. *Asm News*, 68(3).

Selway, C.A., Mills, J.G., Weinstein, P., Skelly, C., Yadav, S., Lowe, A., Breed, M.F. and Weyrich, L.S., 2020. Transfer of environmental microbes to the skin and respiratory tract of humans after urban green space exposure. *Environment International*, 145, p. 106084.

Sharon, G., Segal, D., Ringo, J.M., Hefetz, A., Zilber-Rosenberg, I. and Rosenberg, E., 2010. Commensal bacteria play a role in mating preference of *Drosophila melanogaster*. *Proceedings of the National Academy of Sciences*, 107(46), pp. 20051–20056.

Sheldrake, M., 2020. *Entangled Life: How Fungi Make Our Worlds, Change Our Minds & Shape Our Futures*. New York: Random House.

Shkoporov, A.N., Turkington, C.J. and Hill, C., 2022. Mutualistic interplay between bacteriophages and bacteria in the human gut. *Nature Reviews Microbiology*, pp. 1–13.

Sieradzki, E.T., Ignacio-Espinoza, J.C., Needham, D.M., Fichot, E.B. and Fuhrman, J.A., 2019. Dynamic marine viral infections and major contribution to photosynthetic processes shown by spatiotemporal picoplankton metatranscriptomes. *Nature Communications*, 10(1), pp. 1–9.

Smith, C.J., Emge, J.R., Berzins, K., Lung, L., Khamishon, R., Shah, P., Rodrigues, D.M., Sousa, A.J., Reardon, C., Sherman, P.M. and Barrett, K.E., 2014. Probiotics normalize the gut-brain-microbiota axis in immunodeficient mice. *American Journal of Physiology-Gastrointestinal and Liver Physiology*, 307(8), pp. G793–G802.

Smith, A., 2018. Closed Loop Water Purification System Utilizing an Algae Membrane Photobioreactor for the International Space Station (No. KSC-E-DAA-TN62581).

Song, S.J., Lauber, C., Costello, E.K., Lozupone, C.A., Humphrey, G., Berg-Lyons, D., Caporaso, J.G., Knights, D., Clemente, J.C., Nakielny, S. and Gordon, J.I., 2013. Cohabiting family members share microbiota with one another and with their dogs. *eLife*, 2, p. e00458.

Stefanini, I., Dapporto, L., Legras, J.L., Calabretta, A., Di Paola, M., De Filippo, C., Viola, R., Capretti, P., Polsinelli, M., Turillazzi, S. and Cavalieri, D., 2012. Role of social wasps in *Saccharomyces cerevisiae* ecology and evolution. *Proceedings of the National Academy of Sciences*, 109(33), pp. 13398–13403.

Stein, M.M., Hrusch, C.L., Gozdz, J., Igartua, C., Pivniouk, V., Murray, S.E., Ledford, J.G., Marques dos Santos, M., Anderson, R.L., Metwali, N. and Neilson, J.W., 2016. Innate immunity and asthma risk in Amish and Hutterite farm children. *New England Journal of Medicine*, 375(5), pp. 411–421.

Strachan, D.P., Warner, J., Pickup, J., Schweiger, M.P.H., Stanwell-Smith, R. and Jones, M., 1989. Hygiene hypothesis. *British Medical Journal*, 299, pp. 1259–1260.

Sun, Y., Li, L., Xie, R., Wang, B., Jiang, K. and Cao, H., 2019. Stress triggers flare of inflammatory bowel disease in children and adults. *Frontiers in Pediatrics*, 7, p. 432.

Tessum, C.W., Apte, J.S., Goodkind, A.L., Muller, N.Z., Mullins, K.A., Paolella, D.A., Polasky, S., Springer, N.P., Thakrar, S.K., Marshall, J.D. and Hill, J.D., 2019. Inequity in consumption of goods and services adds to racial–ethnic disparities in air pollution exposure. *Proceedings of the National Academy of Sciences*, 116(13), pp. 6001–6006.

Thomas, H., Ougham, H. and Sanders, D., 2021. Plant blindness and sustainability. *International Journal of Sustainability in Higher Education.*

Thomson, J.A., Widjaja, C., Darmaputra, A.A., Lowe, A., Matheson, M.C., Bennett, C.M., Allen, K., Abramson, M.J., Hosking, C., Hill, D. and Dharmage, S.C., 2010. Early childhood infections and immunisation and the development of allergic disease in particular asthma in a high-risk cohort: a prospective study of allergy-prone children from birth to six years. *Pediatric Allergy and Immunology*, 21(7), pp. 1076–1085.

Timmis, K., Cavicchioli, R., Garcia, J.L., Nogales, B., Chavarría, M., Stein, L., McGenity, T.J., Webster, N., Singh, B.K., Handelsman, J. and de Lorenzo, V., 2019. The urgent need for microbiology literacy in society. *Environmental Microbiology*, 21(5), pp. 1513–1528.

Tomova, A., Bukovsky, I., Rembert, E., Yonas, W., Alwarith, J., Barnard, N.D. and Kahleova, H., 2019. The effects of vegetarian and vegan diets on gut microbiota. *Frontiers in Nutrition*, 6, p. 47.

Traspas, A. and Burchell, M.J., 2021. Tardigrade survival limits in high-speed impacts – implications for panspermia and collection of samples from plumes emitted by ice worlds. *Astrobiology*, 21(7), pp. 845–852.

Troyer, K., 1984. Behavioral acquisition of the hindgut fermentation system by hatchling *Iguana iguana*. *Behavioral Ecology and Sociobiology*, 14(3), pp. 189–193.

Ulrich, K., Becker, R., Behrendt, U., Kube, M. and Ulrich, A., 2020. A comparative analysis of ash leaf-colonizing bacterial communities identifies putative antagonists of *Hymenoscyphus fraxineus*. *Frontiers in Microbiology*, 11, p. 966.

Van Boeckel, T.P., Brower, C., Gilbert, M., Grenfell, B.T., Levin, S.A., Robinson, T.P., Teillant, A. and Laxminarayan, R., 2015. Global trends in antimicrobial use in food animals. *Proceedings of the National Academy of Sciences*, 112(18), pp. 5649–5654.

van de Wouw, M., Boehme, M., Lyte, J.M., Wiley, N., Strain, C., O'Sullivan, O., Clarke, G., Stanton, C., Dinan, T.G. and Cryan, J.F., 2018. Short-chain fatty acids: microbial metabolites that alleviate stress-induced brain–gut axis alterations. *Journal of Physiology*, 596(20), pp. 4923–4944.

van de Wouw, M., Boehme, M., Dinan, T.G. and Cryan, J.F., 2019. Monocyte mobilisation, microbiota & mental illness. *Brain, Behavior, and Immunity*, 81, pp. 74–91.

Waglechner, N., McArthur, A.G. and Wright, G.D., 2019. Phylogenetic reconciliation reveals the natural history of glycopeptide antibiotic biosynthesis and resistance. *Nature Microbiology*, 4(11), pp. 1862–1871.

Walker, M., 2017. *Why We Sleep: The New Science of Sleep and Dreams*. London: Penguin.

Walker, A.R. and Datta, S., 2019. Identification of city specific important bacterial signature for the MetaSUB CAMDA challenge microbiome data. *Biology Direct*, 14(1), pp. 1–16.

Wall Kimmerer, R., 2013. *Braiding Sweetgrass: Indigenous Wisdom, Scientific Knowledge, and the Teachings of Plants*. Minneapolis: Milkweed Editions.

Walsh, D., Bendel, N., Jones, R. and Hanlon, P., 2010. It's not 'just deprivation': why do equally deprived UK cities experience different health outcomes?. *Public health*, 124(9), pp. 487–495.

Wang, S., Ishima, T., Zhang, J., Qu, Y., Chang, L., Pu, Y., Fujita, Y., Tan, Y., Wang, X. and Hashimoto, K., 2020. Ingestion of *Lactobacillus intestinalis* and *Lactobacillus reuteri* causes depression- and anhedonia-like phenotypes in antibiotic-treated mice via the vagus nerve. *Journal of Neuroinflammation*, 17(1), pp. 1–12.

Watkins, H., Robinson, J.M., Breed, M.F., Parker, B. and Weinstein, P., 2020. Microbiome-inspired green infrastructure: a toolkit for multidisciplinary landscape design. *Trends in Biotechnology*, 38(12), pp. 1305–1308.

Westfall, S., Lomis, N. and Prakash, S., 2018. A novel polyphenolic prebiotic and probiotic formulation have synergistic effects on the gut microbiota influencing *Drosophila melanogaster* physiology. *Artificial Cells, Nanomedicine, and Biotechnology*, 46(sup2), pp. 441–455.

Wilson, E.O., 1984. *The Biophilia Hypothesis*. Cambridge, MA: Harvard University Press.

Wilson, E.O., 1992. *The Diversity of Life*. London: Allen Lane. The Penguin Press.

Wittebole, X., De Roock, S. and Opal, S.M., 2014. A historical overview of bacteriophage therapy as an alternative to antibiotics for the treatment of bacterial pathogens. *Virulence*, 5(1), pp. 226–235.

Woo, J. and Lee, C.J., 2020. Sleep-enhancing effects of phytoncide via behavioral, electrophysiological, and molecular modeling approaches. *Experimental Neurobiology*, 29(2), p. 120.

Yong, E., 2016. *I Contain Multitudes: The Microbes Within Us and a Grander View of Life*. New York: Random House.

Zeki, S., Goodenough, O.R. and Sapolsky, R.M., 2004. The frontal cortex and the criminal justice system. *Philosophical Transactions of the Royal Society of London. Series B: Biological Sciences*, 359(1451), pp. 1787–1796.

Zeng, B., Zhao, J., Guo, W., Zhang, S., Hua, Y., Tang, J., Kong, F., Yang, X., Fu, L., Liao, K. and Yu, X., 2017. High-altitude living shapes the skin microbiome in humans and pigs. *Frontiers in Microbiology*, 8, p. 1929.

Zhang, Y., Pechal, J.L., Schmidt, C.J., Jordan, H.R., Wang, W.W., Benbow, M.E., Sze, S.H. and Tarone, A.M., 2019. Machine learning performance in a microbial molecular autopsy context: a cross-sectional postmortem human population study. *PLOS One*, 14(4), p. e0213829.

Zhang, F., Wang, B., Liu, S., Chen, Y., Lin, Y., Liu, Z., Zhang, X. and Yu, B., 2021. *Bacillus subtilis* revives conventional antibiotics against *Staphylococcus aureus* osteomyelitis. *Microbial Cell Factories*, 20(1), pp. 1–15.

Zhu, D., Delgado-Baquerizo, M., Ding, J., Gillings, M.R. and Zhu, Y.G., 2021a. Trophic level drives the host microbiome of soil invertebrates at a continental scale. *Microbiome*, 9(1), pp. 1–16.

Zhu, D., Delgado-Baquerizo, M., Su, J.Q., Ding, J., Li, H., Gillings, M.R., Penuelas, J. and Zhu, Y.G., 2021b. Deciphering potential roles of earthworms in mitigation of antibiotic resistance in the soils from diverse ecosystems. *Environmental Science & Technology*, 55 (11), pp. 7445–7455.

JAKE M. ROBINSON is a microbial ecologist based in the UK. In 2021, he received a PhD from the University of Sheffield. He enjoys researching microbes, ecosystems, social equity issues, and ways to conserve and restore nature. *Invisible Friends* is his first book.

www.jakemrobinson.com

Find Jake on Twitter @_jake_robinson and @MicrobeBook

Index

References to figures appear in *italic* type; those in **bold** type refer to tables.